Bifurcation Theory

In Finite Dimensions

Wm. D. Stone

New Mexico Tech Press
Socorro, New Mexico
http://press.nmt.edu

Copyright ©2013 by William Dean Stone

All rights reserved. Except for fair use in reviews and/or scholarly considerations, no portion of this book may be reproduced in any form without the written permission of the author.

Publisher's Cataloguing-in-Publication Data
Stone, William Dean
 Bifurcation theory: in finite dimensions / Wm. D. Stone.
 iv, 193 p.: ill. ; 28 cm
 Basic examples – Roots of $f(x,a)$ – One-dimensional stability – Period Doubling –Other periods - Annihilation points – Systems of differential equations – Stability of systems – Bifurcation of systems – Hopf bifurcation – Sturm chains – The Routh-Hurwitz criterion – Iterative systems – Modified Routh-Hurwitz –The Brouwer degree – The Rabinowitz theorem.
 ISBN 978-0-9830394-8-8 (pbk.) --
 978-0-9830394-9-5 (ebook)
  1. Bifurcation theory. 2. Differentiable dynamical systems. 3. Functional differential equations.

QA 380.S68 2013
515.625—dc23

OCLC Number: 861760123

This copy printed by CreateSpace, Charleston, SC
Printed in the United States of America

Published by the New Mexico Tech Press, a New Mexico nonprofit organization

Table of Contents

Chapter 1 – Basic Examples 1

Chapter 2 – Roots of $f(x; \alpha)$ 18

Chapter 3 – One-Dimensional Stability 32

Chapter 4 – Period Doubling 52

Chapter 5 – Other Periods 62

Chapter 6 – Annihilation Points 72

Chapter 7 – Systems of Differential Equations 78

Chapter 8 - Stability of Systems 88

Chapter 9 – Bifurcation of Systems 100

Chapter 10 – Hopf Bifurcation 109

Chapter 11 – Sturm Chains 125

Chapter 12 – The Routh-Hurwitz Criterion 135

Chapter 13 – Iterative Systems 147

Chapter 14 – Modified Routh-Hurwitz 157

Chapter 15 – Discrete Hopf Bifurcation 165

Chapter 16 – The Brouwer Degree 175

Chapter 17 – The Rabinowitz Theorem 183

Chapter 1 – Basic Examples

Bifurcation is the study of changes – changes in solutions with changes in parameters.

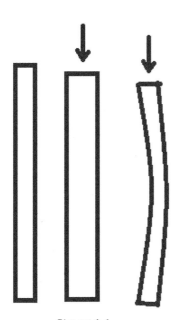

Figure 1.1

The solutions in question may be roots to a polynomial or other types of algebraic equations, solutions (particularly steady-state solutions) to differential equations, or solutions to iterative problems. At certain parameter values the solution may have a qualitative change.

A classic example of this sort of change is that of vertically loading a long, rectangular, elastic beam. Here the parameter is the loading force. For small forces the beam will compress in the vertical

Chapter 1 – Basic Examples

direction, expanding in the horizontal directions, but staying symmetric about the center axis of the beam. Beyond some critical load, however, this will not be what we see. The beam will buckle to one side or the other in a parabolic sort of arc (see Figure 1.1). The symmetric solution is still a solution to the differential equations describing the elastic solid, but it is no longer stable, so it is not what is observed. The point in solution-parameter space where this splitting, or bifurcating, happens is called a **bifurcation point**.

For another example, consider the roots of $f(x;\alpha)$, a polynomial in x with a parameter α.

Example 1.1

$$f(x;\alpha) = x^3 + \alpha x.$$

Solving $f(x;\alpha) = 0$ for x, we find that $x=0$ is a root for all α, and $x = \pm\sqrt{-\alpha}$ is a solution for $\alpha<0$ (we are only interested in real roots. Sketching the roots versus α we get Figure 1.2. This is what we call a **bifurcation diagram**. We can see that at the point ($x=0$; $\alpha=0$) there is a qualitative change in the nature of the solutions, from three real roots to one. This point, (0; 0), is a bifurcation point. This

Chapter 1 – Basic Examples

particular type of bifurcation point, which we will define more carefully later, is sometimes called a pitchfork bifurcation.

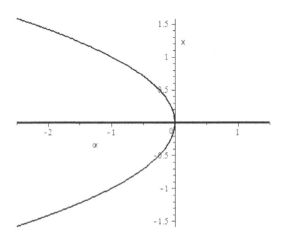

Figure 1.2

Example 1.2

$$f(x; \alpha) = \left(x^2 + \alpha(\alpha - 1)\right)(x - \alpha).$$

Solving for the roots, we get $x = \alpha, x = \pm\sqrt{\alpha - \alpha^2}$. Sketching our bifurcation diagram we get Figure 1.3. Here we can see that we have three points of interest, at ($x=0$; $\alpha=0$), ($x=\frac{1}{2}$; $\alpha=\frac{1}{2}$) and at ($x=0$; $\alpha =1$). The first point, (0; 0), is a pitchfork

Chapter 1 – Basic Examples

bifurcation, the same type as our previous example (although in the other direction). The second point, $(\frac{1}{2};\frac{1}{2})$, is also a bifurcation point. On each side of $\alpha\alpha\alpha = \frac{1}{2}$ $f(x; \alpha)$ has the same number of real roots, so there isn't the same sort of change as at (0; 0), but at $\alpha=\frac{1}{2}$ two of the roots cross. We will see later why this sort of "crossing bifurcation" or **slant bifurcation** is important.

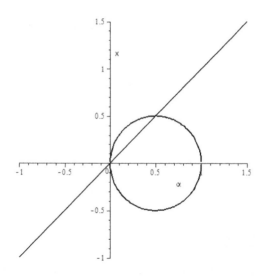

Figure 1.3

The third point, (0; 1), can also be thought of as a bifurcation point, although it is a type we will not work with as much. Two roots come together and disappear. This is sometimes known as an **annihilation point**.

For these simple examples we were able to see the roots without difficulty. For more sophisticated problems we could use the built-in numerical solvers in programs such as Maple© or Matlab©, or a numeric scheme such as Newton's method. For Newton's method, let x_0 be a first guess at a root, then $x_1 = x_0 + \varepsilon$, with ε unknown. We want $f(x_1) = f(x_0 + \varepsilon) = 0$. Using a linear Taylor's polynomial to approximate $f(x_0 + \varepsilon)$ gives

$$f(x_0 + \varepsilon) \cong f(x_0) + \varepsilon \frac{df}{dx}(x_0) = 0.$$

Solving for ε gives

$$\varepsilon = -\frac{f(x_0)}{f'(x_0)}$$

and thus

$$x_1 = x_0 - \frac{f(x_0)}{f'(x_0)}.$$

Continuing we have

Chapter 1 – Basic Examples

$$x_{n+1} = x_n - \frac{f(x_n)}{f'(x_n)},$$

And the sequence $\{x_n\}$ will usually converge to a root of f.

Example 1.3

Consider the first order differential equation

$$\frac{dx}{dt} = x(\alpha - x - 1).$$

Solving by separation of variables we get

$$\ln\left|\frac{x}{x - \alpha + 1}\right| = (\alpha - 1)t + C$$

or

$$\frac{x}{x - \alpha + 1} = Ae^{(\alpha-1)t}.$$

Solving for x, we get

$$x(t) = \frac{(\alpha - 1)Ae^{(\alpha-1)t}}{1 + Ae^{(\alpha-1)t}}.$$

Chapter 1 – Basic Examples

Clearly the solution depends on α and the initial condition. However, consider the limit as t approaches ∞. If $\alpha<1$ then $\lim_{t\to\infty} e^{(\alpha-1)t} = 0$, so $\lim_{t\to\infty} x(t) = 0$. If $\alpha>1$, $\lim_{t\to\infty} e^{(\alpha-1)t} = \infty$, so $\lim_{t\to\infty} x(t) = \alpha - 1$. See Figure 1.4. There is a change in the asymptotically stable steady state at ($x=0$; $\alpha=1$). By solving for $\frac{dx}{dt} = 0$, we see that $x=0$ and

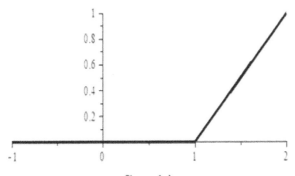

Figure 1.4

$x=\alpha-1$ are always steady states. At $\alpha=1$ the two steady states cross and switch stability as shown in Figure 1.5, the bifurcation diagram. This is what is called an **exchange of stability** bifurcation.

We will also be considering iterative problems. For instance, consider the function $f(x) = \frac{3}{2}x - x^2$. Given an x_0, we can define a sequence $\{x_n\}$ by $x_{k+1} = f(x_k)$. As an example, let $x_0 = 0.1$. Then $x_1 = f(x_0) = \frac{3}{2}x_0 - x_0^2 =$

Chapter 1 – Basic Examples

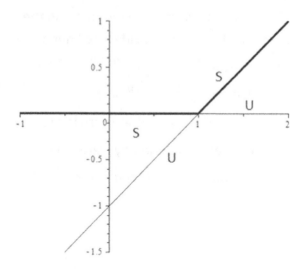

Figure 1.5

$0.14.; x_2 = f(x_1) = .1904; ...$ If we keep going, we see that the sequence appears to settle down to 0.5. Try another starting value: $x_0 = 0.9, x_1 = 0.54, x_2 = 0.5184, ...$ Again, the sequence appears to converge to a constant value of 0.5.

If we were to start at $x_0 = 0.5, x_1 = f(x_0) = 0.5$, so $x_0 = x_1 = x_2 = \cdots = 0.5$; the sequence is constant. A value x^* such that $f(x^*) = x^*$ is called a **fixed point** of f. Fixed points for iterations are much like steady states for differential equations: constant solutions.

For sequences defined this way,

Chapter 1 – Basic Examples

$$x_1 = f(x_0), x_2 = f(x_{10}) = f \circ f(x_0), \ldots, x_k = f(x_{k-1}) = \frac{f \circ f \circ f \circ \ldots \circ f(x_0)}{(k \text{ times})}.$$

For simpler notation we will define $f^{\circ n}$ as f composed with itself n times. Thus for our iterative problem we have $x_k = f^{\circ k}(x_0)$. These sequences are often difficult to get in closed form, but are very easy to generate numerically, and are used as models in many fields.

The formula we obtained, $x_k = f^{\circ k}(x_0)$, is a solution, but as we will see in the exercises, it is not very useful. An interesting (and amusing!) graphical approach is the **web diagram**.

Chapter 1 – Basic Examples

To make a web diagram, start by graphing y=f(x), the function that generates your sequence. On the same graph, plot y=x. As an example we will use $f(x) = \frac{3}{2}x - x^2$ (see Figure 1.6). Start at the point $(x_0, 0)$. Move vertically until you hit the graph of f(x). This will be the point $(x_0, f(x_0))$ which is (x_0, x_1). Now move horizontally to the line y=x. The second coordinate will not have changed, so you are at (x_1, x_1). Now moving vertically to y=f(x) puts you at $(x_1, f(x_1))$ or (x_1, x_2). Continuing in this manner (vertically to the curve, horizontally to the line) we generate our sequence,

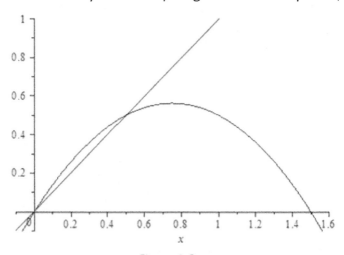

Figure 1.6

as shown in Figure 1.7.

Chapter 1 – Basic Examples

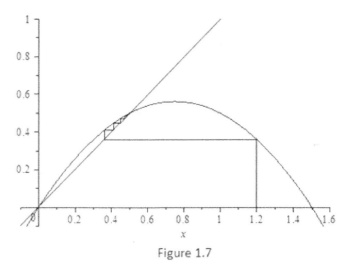

Figure 1.7

Note where $y=f(x)$ and $y=x$ intersect. This will be the solutions of $f(x)=x$, which are the fixed points of f. In our example we see that we have two fixed points, one at $x=0$ and the other at $x=\frac{1}{2}$. Using the web diagram we can see that sequences starting near $x=\frac{1}{2}$ approach $\frac{1}{2}$, while sequences starting close to 0 (but not exactly at 0) go away from 0. The fixed point at $\frac{1}{2}$ is stable; that at 0 is unstable.

Chapter 1 – Basic Examples

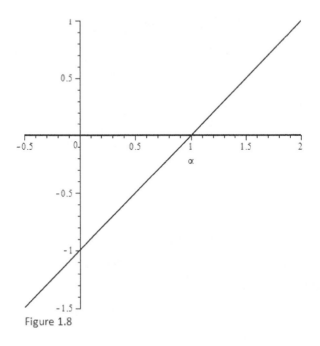

Figure 1.8

If we have a family of functions with a parameter α, we can again consider changes in the fixed points, and the long term behavior of the sequences, with changes in α. We have all the possibilities we have already seen: slant bifurcation/exchange of stability, pitchfork bifurcation, annihilation point, plus some other possibilities we will see later. As before, we will be iterating the function f. If $x_1 = f(x_0; \alpha)$, then $x_2 =$

Chapter 1 – Basic Examples

$f(x_1; \alpha) = f(f(x_0; \alpha); \alpha)$. The notation we will use for this is $x_2 = f^{\circ 2}(x_0; \alpha)$, or in general $f^{\circ k}(x; \alpha) = f(f^{\circ k-1}(x; \alpha); \alpha)$.

As an example, consider $f(x; \alpha) = \alpha x - x^2$ (Note that $f(x; \frac{3}{2})$ is the problem we were just considering). Solving $\alpha x - x^2 = x$ we find that we have two fixed points, 0 and α-1 (see Figure 1.8). Previously, we saw that at $\alpha = \frac{3}{2}$ the fixed point at $x=\frac{1}{2}$ ($x=\alpha$-1) was stable and the one at x=0 was unstable. If we try $\alpha=\frac{1}{2}$ we get the web diagram in Figure 1.9.

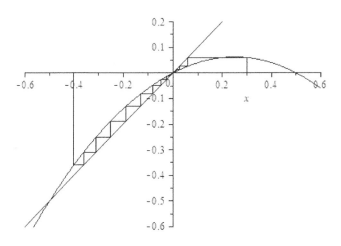

Figure 1.9

Here we see that $x=0$ is stable, and the fixed point at $x=-\frac{1}{2}$ ($x=\alpha-1$) is unstable. We might suspect (and will later prove) a bifurcation diagram like the one shown in Figure 1.10, an exchange of stability bifurcation point.

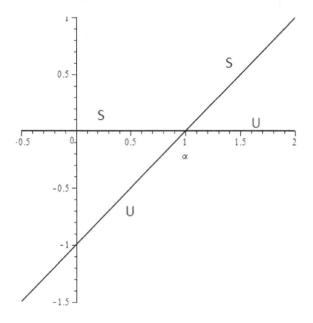

Figure 1.10

Chapter 1 – Basic Examples

Chapter 1 Exercises

1.1 Sketch the real roots of f versus α for the following functions. Identify the bifurcation points as pitchfork bifurcation points, slant bifurcation points or annihilation points.

a. $f(x; \alpha) = x(x - \alpha)(x^2 + \alpha + 1)$

b. $f(x; \alpha) = x - \alpha x(1 - x)$

c. $f(x; \alpha) = x^4 - \alpha x^3 + 1$ (Solve these last two numerically if you want.)

d. $f(x; \alpha) = e^{\alpha x} - x^2$

1.2 For the following problems, find and sketch the steady states versus α. Using a numeric differential equation solver, solve for several αs with several different initial values each. What do you observe?

a. $\frac{dx}{dt} = x(\alpha - x)$

b. $\frac{dx}{dt} = x^3 - \alpha x$

c. $\frac{d^2x}{dt^2} + (1 - \alpha)\frac{dx}{dt} + x(x - \alpha) = 0$

Chapter 1 – Basic Examples

1.3 Consider the example in the book, the sequence generated by $x_{k+1} = \frac{3}{2}x_k - x_k^2$. Draw the web diagram and determine how the $\lim_{n\to\infty} x_n$ depends on x_0.

1.4 For $f(x; \alpha) = \alpha x - x^2$, numerically calculate $\{x_n\}_{n=1}^{200}$ with x_0 chosen randomly, for $\alpha = 0$, .05, .1, .15, ..., 3.5. (Spreadsheets are an easy way to do this.) What do you observe? Careful!

1.5 Find fixed points, choose an appropriate range for α, then repeat Exercise 1.4 for $f(x; \alpha) = x^2 - \alpha$.

Chapter 1 – Basic Examples

Chapter 2 – Roots of $f(x; \alpha)$

The first problems we will look at are the roots of functions. In particular we will consider the case where we know one root, which continues throughout our interval of interest for α. That is, we have x*(α) such that f(x*(α); α) =0 for all α in some open interval, say (α_1, α_2). Note that this will not cover the case of annihilation points.

For simplicity's sake, it is easier to assume that x*(α)=0 for all α. This is just a shift of our problem. Given $\hat{f}(x; \alpha)$ and x*(α) such that $\hat{f}(x^*(\alpha); \alpha) = 0$, define y = x-x*(α) and consider f(y; α) defined to be $\hat{f}(y + x^*(\alpha); \alpha)$. Then $f(0; \alpha) = \hat{f}(x^*(\alpha); \alpha) = 0$; y=0 is equivalent to x=x*(α). Note also that $\frac{\partial f}{\partial y}(y; \alpha) = \frac{\partial \hat{f}}{\partial y}(y + x^*(\alpha); \alpha) = \frac{\partial \hat{f}}{\partial x}(x; \alpha)$. Thus we can assume f(0; α)=0 for all α, without loss of generality.

We also require f to be continuously differentiable in the α-interval of interest, and in a neighborhood of x=0, i.e., $f \in C^1[(-\varepsilon, \varepsilon) \times$

Chapter 2 – Roots of f(x; α)

(α_1, α_2)]. Under this condition we can use Taylor's Theorem to expand f in x.

Taylor's Theorem

If $g(x) \in C^k$ in a neighborhood of x_0, then for x in a (possibly smaller) neighborhood
$$g(x) = g(x_0) + \frac{dg}{dx}(x_0) \cdot (x - x_0) + \frac{1}{2}\frac{d^2g}{dx^2}(x_0) \cdot (x - x_0)^2 + \cdots + \frac{1}{k!}\frac{d^kg}{dx^k}(x_0) \cdot (x - x_0)^k + R_k(x, x_0)$$
where $R_k(x, x_0)$ is $o(x - x_0)^k$. If $g \in C^{k+1}$ in the neighborhood, R_k is $O(x - x_0)^{k+1}$.

The notation $o(x - x_0)^k$ is read "little oh of $(x - x_0)^k$"; $O(x - x_0)^k$ is read either "big oh of $(x - x_0)^k$" or "order $(x - x_0)^k$". A function P(x) is $o(Q(x))$ at $x = x_0$ if $\lim_{x \to x_0} \frac{P(x)}{Q(x)} = 0$. For instance, $x \ln(x)$ is $o(x^a)$ for any $a>1$.

We say that M(x) is O(G(x)) at $x = x_0$ if $\lim_{x \to x_0} \left|\frac{M(x)}{G(x)}\right| = c < \infty$. For instance, $\frac{x^3+x^2}{x+2} = O(x^2)$ at x=0. (The functions we compare to, called the **gauge functions**, are often powers of x but can be other sets of functions.) Since 0 is a constant less than infinity, any function that is o(Q) is also O(Q); o(Q) is the stronger condition. However, $o(x^k)$ is a

19

Chapter 2 – Roots of f(x; α)

weaker condition than $O(x^{k+1})$. For example, $x^{3/2}$ is $o(x^1)$ but not $O(x^2)$. Since we are usually concerned with behavior near zero, we will usually be looking at o or O at $x=0$. We may not always specifically state this; it should be clear from context.

Going back to our problem, we are assuming that $f(x; \alpha) \in C^1$ so we have

$$f(x; \alpha) = f(0; \alpha) + \frac{\partial f}{\partial x}(0; \alpha)x + R_1(x; \alpha).$$

Under our assumptions, $f(0; \alpha)=0$. $\frac{\partial f}{\partial x}(0; \alpha)$ is a function of α only; we will call it $\lambda(\alpha)$. The remainder term, $R_1(x; \alpha)$, is $o(x)$ so we can write it as $x \cdot h(x; \alpha)$ where h is $o(1)$. Now we have

$$f(x; \alpha) = \lambda(\alpha)x + x \cdot h(x; \alpha) \qquad (2.1)$$

where $\lim_{x \to 0} h(x; \alpha) = 0$ for all α.

Next we will use the Implicit Function Theorem to further investigate the roots of f.

Implicit Function Theorem (IFT)
Let V_1 and V_2 be vector spaces, with W an open set in $V_1 \times V_2$. Let $F: W \to V_1$ be C^1.

Chapter 2 – Roots of f(x; α)

Suppose $(x_0, y_0) \in W$ is such that $F(x_0, y_0) = c_0$ and the linear operator $\frac{\partial F}{\partial x}\big|_{(x_0, y_0)} : V_1 \to V_1$ is invertible. Then there are open sets $U_1 \subset V_1, U_2 \subset V_2, (x_0, y_0) \in U_1 \times U_2 \subset W$ and a unique C^1 map $X: U_2 \to U_1$ such that $X(y_0) = x_0$ and $F(X(y), y) = c_0$ for all y in a neighborhood of y_0. Further, for $(x, y) \in U_1 \times U_2$, if $x \neq X(y)$ then $F(x, y) \neq c_0$. The proof can be found in any standard Real Analysis text.

In other words, if F has a non-singular derivative at a point, there is a unique level curve (or higher dimensional analog) to F through the point.

For our problem, $f(x; \alpha) = \lambda(\alpha)x + xh(x; \alpha)$, we know that $f(0; \alpha) = 0$ for all α. By the Implicit Function Theorem, if $\frac{\partial f}{\partial x}(0; \alpha_0)$ is invertible there is a unique curve through the point $(0; \alpha_0)$ that satisfies the equation f(x; α)=0, namely the solution we already have, x=0. For our $f, \frac{\partial f}{\partial x} = \lambda(\alpha) + h(x; \alpha) + x\frac{\partial h}{\partial x}(x; \alpha)$, and $\frac{\partial f}{\partial x}(0; \alpha_0) = \lambda(\alpha_0) + h(0; \alpha_0) = \lambda(\alpha_0)$. (Recall

Chapter 2 – Roots of f(x; α)

that $h(0; \alpha)=0$ for all α.) Thus if $\lambda(\alpha_0) \neq 0$, $x = 0$ is the unique solution through the point $(0; \alpha_0)$.

What about the points $(0; \alpha_1)$ such that $\lambda(\alpha_1)=0$? These points, where the Implicit Function Theorem fails to guarantee a unique solution, are possible bifurcation points.

We are interested in roots of $f(x; \alpha) = 0$ other than x=0. Dividing out the root we know, we are looking for solutions to

$$\lambda(\alpha) + h(x; \alpha) = 0. \qquad (2.2)$$

We already know a point that satisfies (2.2). Since $h(0; \alpha)=0$ for all α, and $\lambda(\alpha_1)=0$, the point $(0; \alpha_1)$ satisfies the equation. Can we use the Implicit Function Theorem to extend this to a solution curve?

Let us define $g(x; \alpha) = \lambda(\alpha) + h(x; \alpha)$; we know that $g(x; \alpha_1) = 0$. To apply the IFT we consider $\frac{\partial g}{\partial x} = \frac{\partial h}{\partial x}$. Unfortunately, all we know about h is that it is o(1) at x=0; $\frac{\partial h}{\partial x}$ may be non-zero at $(0; \alpha_1)$, but it also may be zero, or h may not even be differentiable at that point.

It turns out that in this case we know more about the partial of g (and of h) with respect to α:

Chapter 2 – Roots of f(x; α)

$\frac{\partial g}{\partial \alpha}\big|_{(0;\alpha_1)} = \frac{d\lambda}{d\alpha}(\alpha_1) + \frac{\partial h}{\partial \alpha}(0;\alpha_1)$. Since we are taking the partial with respect to α, we can substitute $x=0$ first. Thus $\frac{\partial h}{\partial \alpha}(0;\alpha_1) = \frac{d}{d\alpha}(h(0;\alpha_1)) = \frac{d}{d\alpha}(0) = 0$. This gives us $\frac{\partial g}{\partial \alpha}\big|_{(0;\alpha_1)} = \frac{d\lambda}{d\alpha}(\alpha_1)$, and if $\lambda'(\alpha_1) \neq 0$, the Implicit Function Theorem implies that there is a function $A(x)$, defined in some neighborhood of $x=0$, such that $A(0)=\alpha_1$ and $g(x;A(x))=0$. Since this function is defined in a neighborhood of $x=0$, it is definitely not the solution $x \equiv 0$ that we already had, although it intersects the previously known solution at $(0;\alpha_1)$; $(0;\alpha_1)$ is a bifurcation point.

What we have developed is the First Bifurcation Theorem.

First Bifurcation Theorem

Suppose $f: \mathbb{R} \times \mathbb{R} \to \mathbb{R}$ is such that $f \in C^1([u,v] \times [a,b])$ and there exists a continuous function $x^*: [a,b] \to [u,v]$ such that $f(x^*(\alpha); \alpha) = 0$. Let $\frac{\partial f}{\partial x}(x^*(\alpha); \alpha) = \lambda(\alpha)$.

    i)   If $\lambda(\alpha_0) \neq 0$, then there is an open ball about $(x^*(\alpha_0); \alpha_0)$ that contains no points

Chapter 2 – Roots of f(x; α)

ii) satisfying $f(x;\alpha) = 0$ other than the points on the curve $(x^*(\alpha);\alpha)$.

If $\lambda(\alpha_1) = 0$ and $\frac{d\lambda}{d\alpha}(\alpha_1) \neq 0$ then there exists a unique curve of the form $(x; A(x))$ such that $f(x; A(x))=0$. The two solutions intersect at $(x_1;\alpha_1)$, i.e., $A(x_1) = \alpha_1$, and $x^*(\alpha_1) = x_1$; the point $(x_1;\alpha_1)$ is a bifurcation point.

If f is sufficiently smooth, we can find an expansion of the bifurcated solution. Consider $f(x;\alpha) = \lambda(\alpha)x + xh(x;\alpha)$. We are interested in the solution other than x=0; we want a solution to Equation 2.2 $\lambda(\alpha) + h(x;\alpha) = 0$.

If our bifurcation point is $(0;\alpha_0)$, we are looking for a solution of the form $(x; A(x))$, near x=0, such that $A(0)=\alpha_0$. Substituting $\{x = \varepsilon, \alpha = \alpha_0 + \varepsilon\alpha_1 + \varepsilon^2\alpha_2 + \varepsilon^3\alpha_3 + O(\varepsilon^4)\}$ into Equation 2.2 and expanding in ε we get

$$(\lambda(\alpha_0) + h(0;\alpha_0)) + \varepsilon\left(\frac{d\lambda}{d\alpha}(\alpha_0)\alpha_1 + \frac{\partial h}{\partial x}(0;\alpha_0) + \frac{\partial h}{\partial \alpha}(0;\alpha_0)\alpha_1\right) + O(\varepsilon^2). \qquad (2.3)$$

Chapter 2 – Roots of f(x; α)

Since we know that $h(0; \alpha) = 0$ for all α, and that $\lambda(\alpha_0) = 0$, we can simplify this to

$$\varepsilon \left(\frac{d\lambda}{d\alpha}(\alpha_0)\alpha_1 + \frac{\partial h}{\partial x}(0; \alpha_0) \right) + \varepsilon^2 \left(\frac{d\lambda}{d\alpha}(\alpha_0)\alpha_2 + \frac{1}{2}\frac{d^2\lambda}{d\alpha^2}(\alpha_0)\alpha_1^2 + \frac{\partial^2 h}{\partial x^2}(0; \alpha_0) \right) + O(\varepsilon^3). \quad (2.4)$$

In the $O(\varepsilon)$ term, the only unknown is α_1. As long as $\frac{d\lambda}{d\alpha}(\alpha_0) \neq 0$, we can proceed to solve for $\alpha_1, \alpha_2, \alpha_3$, and so on for as many terms as we like. Note that if $\frac{d\lambda}{d\alpha}(\alpha_0) = 0$ Equation 2.4 may or may not be solvable, depending on $\frac{\partial h}{\partial x}(0; \alpha_0)$.

As an example, consider the function $f(x; \alpha) = e^x - (x + x^3)\alpha - 1$. We can see that x=0 is a solution. $\frac{\partial f}{\partial x}(0; \alpha) = 1 - \alpha$ so this is $\lambda(\alpha)$. Note that $\lambda(1) = 0$, so $x=0, \alpha = 1$ is our bifurcation point. Now we can write f as $(1 - \alpha)x + x\left(\frac{e^x - 1}{x} - 1 - \alpha x^3\right)$. Note that h, $\left(\frac{e^x - 1}{x} - 1 - \alpha x^3\right)$, approaches 0 at x=0.

We are looking for a solution to $\lambda(\alpha) + h(x; \alpha) = 0$, that is

Chapter 2 – Roots of f(x; α)

$$\left(\frac{e^x-1}{x} - 1 - \alpha x^3\right) = 0 \quad (1-\alpha) +$$

(2.5)

near $x=0$, $\alpha = 1$. Substituting $\{x = \varepsilon, \alpha = 1 + \varepsilon\alpha_1 + \varepsilon^2\alpha_2 + \varepsilon^3\alpha_3 + O(\varepsilon^4)\}$ into Equation 2.5 and taking a Taylor's expansion in ε, we get $\left(-\alpha_1 + \frac{1}{2}\right)\varepsilon + \left(-\alpha_2 - \frac{5}{6}\right)\varepsilon^2 + O(\varepsilon^3)$, so near $x=0$ our bifurcated solution is given by $\left(x; 1 + \frac{x}{2} - \frac{5x^2}{6} + O(x^3)\right)$. (See Figure 2.1.)

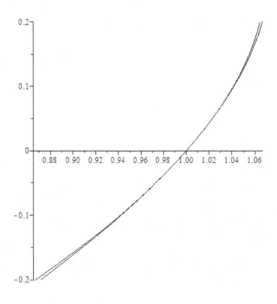

Figure 2.1

Chapter 2 – Roots of f(x; α)

For another example, consider the function $g(x; \alpha) = \alpha x^2 + \alpha x - x^3$. Again we see that $x=0$ is a solution to $g(x; \alpha) = 0$ for all α. $\frac{\partial g}{\partial x}(0; \alpha) = \lambda(\alpha) = \alpha$, so our bifurcation point is at $(x; \alpha) = (0; 0)$. Factoring off the known solution, we are left with

$$\lambda(\alpha) + h(x; \alpha) = \alpha + (\alpha x - x^2) = 0. \qquad (2.6)$$

Assuming x small, we substitute $\alpha = 0 + \alpha_1 x + a_2 x^2 + a_3 x^3 + \cdots$ into Equation 2.6 and expand. This gives $(\alpha_1)x + (\alpha_1 - 1 + \alpha_2)x^2 + (\alpha_2 + \alpha_3)x^3 + O(x^4)$ so we find $\alpha_1 = 0, \alpha_2 = 1, \alpha_3 = -1$, and we find our bifurcated solution is approximately $(x; x^2 - x^3 + O(x^4))$. (See Figure 2.2.)

Chapter 2 – Roots of f(x; α)

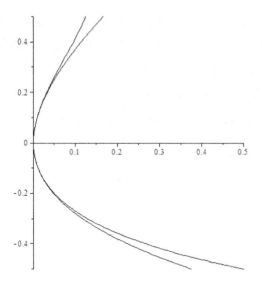

Figure 2.2

Chapter 2 – Roots of f(x; α)

Chapter 2 Exercises

2.1 Let $f(x; \alpha) = (x + \alpha^2)(x^2 + \alpha)$.

a) Find and graph the solutions to $f(x; \alpha) = 0$.

b) Let $x = x^*(\alpha)$ be the solution that exists for all α. Define $y = x - x^*(\alpha)$ and substitute into f to eliminate x.

c) Find and graph the solutions to $\tilde{f}(y; \alpha) = 0$.

d) Compare the graphs in parts a & c.

e) Identify any bifurcation points and their type.

2.2 For the following functions, find and graph the solutions to $f(x; \alpha) = 0$. Identify all bifurcation points and their types.

a) $f(x; \alpha) = (x - \alpha)(\alpha^2 - x)$

b) $f(x; \alpha) = x^2 - x(\alpha - 1)^2$

c) $f(x; \alpha) = x^3 - x + \alpha$

d) $f(x; \alpha) = \alpha x^2 + x - 1$

Chapter 2 – Roots of f(x; α)

2.3 For the following functions, find $\lambda(\alpha)$ as defined in this chapter. Where does $\lambda(\alpha) = 0$? Now plot solutions to $f(x; \alpha) = 0$.

a) $f(x; \alpha) = x^3 - x^2 - \alpha x$

b) $f(x; \alpha) = x^2 - 2x + 2\alpha - \alpha^2$

c) Repeat the problem in part b, using the other root as x^*.

2.4 For the following functions, find $\lambda(\alpha)$ as defined in this chapter. Where does $\lambda(\alpha) = 0$? Does the first bifurcation theorem apply? Now plot solutions to $f(x; \alpha) = 0$. Explain.

a) $f(x; \alpha) = x(x - \alpha^2)$

b) $f(x; \alpha) = x(x^2 + \alpha^2)$

c) $f(x; \alpha) = x(x^4 - \alpha^4)$

d) $f(x; \alpha) = x(x - \alpha^3)$

2.5 For the following functions, find the bifurcation point and expand the bifurcated solution.

a) $f(x; \alpha) = \sin(\alpha x) - x$

b) $f(x; \alpha) = (3 - \alpha)x^2 + (\alpha - 2)x$

Chapter 2 – Roots of f(x; α)

c) $f(x; \alpha) = x^3 - \alpha x^2 - (\alpha + 1)x + \alpha^2 + \alpha$ (NB: $x = \alpha$ works.)

Chapter 3 – One-Dimensional Stability

In the previous chapter we were considering the roots of a function. One place where this comes up is in finding the steady states of a differential equation such as

$$\frac{dx}{dt} = f(x; \alpha). \tag{3.1}$$

Another situation would be finding the fixed points of an iterative problem such as

$$x_{n+1} = g(x_n; \alpha), \tag{3.2}$$

although in this instance we would be looking for roots of $g(x; \alpha) - x$. In these cases, however, there is more to the problem. As an example consider

$$\frac{dx}{dt} = \alpha - x - \alpha x^2 + x^3. \tag{3.3}$$

Factoring we get

$$\frac{dx}{dt} = (x+1)(x-1)(x-\alpha).$$

Chapter 3 – One-Dimensional Stability

We could separate variables and solve, (getting $\frac{(x+1)^{\alpha-1}(x-\alpha)^2}{(x-1)^{\alpha+1}} = Ce^{2(\alpha^2-1)t}$), but the exact solution is not easy to interpret. Instead we will **linearize** around each steady state. First consider x close to 1. Let $x(t) = 1 + \varepsilon x_1(t)$, where ε is small. We will substitute this into Equation 3.3 and expand in powers of ε. This gives

$$\varepsilon \frac{dx_1}{dt} = \alpha - (1 + \varepsilon x_1) - \alpha(1 + \varepsilon x_1)^2 + (1 + \varepsilon x_1)^3 \Rightarrow$$

$$\varepsilon \frac{dx_1}{dt} = (\alpha - 1 - \alpha + 1) + \varepsilon(-x_1 - 2\alpha x_1 + 3x_1) + O(\varepsilon^2)$$

so

$$\frac{dx_1}{dt} = 2(1-\alpha)x_1 + O(\varepsilon). \tag{3.4}$$

Neglecting small ($O(\varepsilon)$) terms, we get that $x_1(t) = Ce^{2(1-\alpha)t}$, and $x(t) = 1 + \varepsilon Ce^{2(1-\alpha)t} + O(\varepsilon^2)$. If $\alpha > 1$, this solution approaches 1. (If $\alpha < 1$, we can say the solution grows away from 1, but since nonlinear terms would then become important, we can't say exactly what the solution does.) Thus for $\alpha < 1$ the steady state x=1 is unstable, and for $\alpha > 1$ it is stable.

Chapter 3 – One-Dimensional Stability

Similarly, if we consider $x = -1 + \varepsilon x_1$ we get

$$\frac{dx_1}{dt} = 2(1 + \alpha)x_1 + O(\varepsilon).$$

Setting $x = \alpha + \varepsilon x_1$ gives

$$\frac{dx_1}{dt} = (\alpha^2 - 1)x_1 + O(\varepsilon).$$

Thus x=-1 is stable for $\alpha < -1$ and otherwise unstable; $x = \alpha$ is stable for $-1 < \alpha < 1$ and otherwise unstable. Putting this all together we obtain the bifurcation diagram. (See Figure 3.1.)

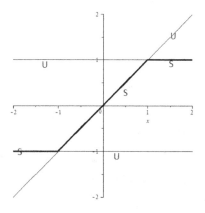

Figure 3.1

In general, we have

$$\frac{dx}{dt} = f(x; \alpha) \qquad (3.5)$$

Chapter 3 – One-Dimensional Stability

with a steady state $x = x_0(\alpha)$. Substitute $x = x_0(\alpha) + \varepsilon x_1(t; \alpha)$ into Equation 3.5 and expand in powers of ε. If f is at least C^1 we get

$$\varepsilon \frac{dx_1}{dt} = f(x_0(\alpha); \alpha) + \varepsilon \frac{\partial f}{\partial x}(x_0(\alpha); \alpha)x_1(t; \alpha) + o(\varepsilon).$$

(If f is C^2 the last term is $O(\varepsilon^2)$.) Since x_0 is a steady state, $f(x_0(\alpha); \alpha) = 0$. Using this, dividing by ε, and letting ε approach zero gives

$$\frac{dx_1}{dt} = \frac{\partial f}{\partial x}(x_0(\alpha); \alpha)x_1$$

or, using the notation we defined earlier

$$\frac{dx_1}{dt} \cong \lambda(\alpha)x_1. \tag{3.6}$$

Solving this we get $x_1 \cong Ce^{\lambda(\alpha)t}$, so

$$x(t) \cong x_0(\alpha) + C\varepsilon e^{\lambda(\alpha)t}$$

for x close to $x_0(\alpha)$. If $\lambda(\alpha) < 0, x = x_0(\alpha)$ is **asymptotically stable**, i.e., solutions that start close to $x_0(\alpha)$ approach $x_0(\alpha)$ as t increases. If $\lambda(\alpha) > 0$ solutions that start close to $x_0(\alpha)$ (but not exactly on it) move away from $x_0(\alpha)$: $x=x_0(\alpha)$ is unstable.

Chapter 3 – One-Dimensional Stability

The stability of $x = x_0(\alpha)$ changes at those values α_0 of α such that $\lambda(\alpha_0) = 0$. Recall that these were also the points where the Implicit Function Theorem failed to guarantee a unique solution, and if $\frac{d\lambda}{d\alpha} \neq 0$, we were guaranteed another solution curve, $(x; A(x))$ through the point $(x_0(\alpha_0); \alpha_0)$. Thus stability is changing at the bifurcation points.

In our example, when two solutions crossed and one lost stability, the other solution became stable (see Figure 3.1). This sort of bifurcation point is called an **exchange of stability** point.

We are interested in the stability of the bifurcated solution $(x; A(x))$. To determine stability we need to find the sign of $\frac{\partial f}{\partial x}(x; A(x))$ near $x=0$. Differentiating (assuming f is sufficiently smooth to allow us to do so) gives

$$\frac{\partial}{\partial x} x\lambda(\alpha) + xh(x; \alpha)|_{(x;A(x))} = \lambda(A(x)) + h(x; A(x)) + xh_x(x; A(x)). \qquad (3.7)$$

We don't know very much about $h_x(x; A(x))$, except that

$$\lambda(A(x)) + h(x; A(x)) = 0, \qquad (3.8)$$

Chapter 3 – One-Dimensional Stability

so

$$\left(\frac{d\lambda}{dx}(A(x)) + h_\alpha(x; A(x))\right)\frac{dA}{dx} + h_x(x; A(x)) = 0.$$
(3.9)

Substituting Equations 3.8 and 3.9 into 3.7 we get

$$\left.\frac{\partial f}{\partial x}\right|_{(x;A(x))} = -x\frac{dA}{dx}\left(\frac{d\lambda}{dx}(A(x)) + h_\alpha(x; A(x))\right).$$
(3.10)

If x is near zero, (near the bifurcation point), then $A(x)$ is near α_0, and $\frac{d\lambda}{dx}(A(x)) + h_\alpha(x; A(x))$ will have the same sign as $\frac{d\lambda}{dx}(\alpha_0) + h_\alpha(0; \alpha_0) = \frac{d\lambda}{dx}(\alpha_0)$, which by our assumption is not zero. To make the arithmetic easier, let us assume $\frac{d\lambda}{dx}(\alpha_0) > 0$, so that the base equation loses stability at α_0. Then, $\left.\frac{\partial f}{\partial x}\right|_{(x;A(x))}$ has the same sign as $-x\frac{dA}{dx}$ near the bifurcation point. Again, near the bifurcation point, $\frac{dA}{dx}$ is close to $\frac{A(x)-A(0)}{x-0} = \frac{A(x)-\alpha_0}{x}$. Thus, $\left.\frac{\partial f}{\partial x}\right|_{(x;A(x))}$ has the same sign as $\alpha_0 - \alpha$. Therefore, $(x; A(x))$ near $(0; \alpha_0)$ is stable precisely where $x=0$ is unstable.

Chapter 3 – One-Dimensional Stability

What we have just shown is the stability theorem for continuous, type-one (exchange of stability) bifurcation points:

Continuous Type-one Stability Theorem

Suppose $f(x; \alpha)$, $\lambda(\alpha)$, and the bifurcation point $(x_1; \alpha_1)$ are all as given in the First Bifurcation Theorem (Chapter 2) with the additional requirement that f be at least C^2 in a neighborhood of $(x_1; \alpha_1)$. Then for α sufficiently close to α_1, if the bifurcated solution exists and the original solution is unstable, the bifurcated solution is stable; if the bifurcated solution exists and the original solution is stable, the bifurcated solution is unstable. (The bifurcated solution may only exist on one side of α_1, for instance with pitchfork bifurcations. If two branches appear on the same side, they will both have the opposite stability of the original solution.)

As an example consider

Chapter 3 – One-Dimensional Stability

$$\frac{dx}{dt} = x^3 + \alpha x^2 + \alpha^2 x - \alpha x. \qquad (3.11)$$

We see that x=0 is a solution for all α. Taking $\frac{\partial f}{\partial x}$ we find that $\lambda(\alpha) = \alpha^2 - \alpha$, so x=0 is stable for $0 < \alpha < 1$. Thus we have two possible bifurcation points, (0; 0) and (0; 1).

Next we want to find the bifurcated solutions. Since f is C^∞, $A(x)$ is also. Looking near (0; 0) we want a solution to

$$\alpha^2 - \alpha + (\alpha x + x^2) = 0 \qquad (3.12)$$

near x=0. Let $A(x) = 0 + a_1 x + a_2 x^2 + \cdots$ and substitute into Equation 3.12. We get

$$a_1^2 x^2 + 2a_1 a_2 x^3 + \cdots - a_1 x - a_2 x^2 - \cdots + a_1 x^2 + a_2 x^3 + \cdots + x^2 = 0$$

or

$$(-a_1)x + (a_1^2 - a_2 + a_1 + 1)x^2 \\ + (2a_1 a_2 - a_3 + a_2)x^3 + O(x^4) \\ = 0.$$

Setting each coefficient equal to zero, we find that $a_1 = 0, a_2 = 1, a_3 = 1$, and thus $A(x) = x^2 + x^3 + O(x^4)$. For x near 0, A(x)>0. Since this is where x=0 is stable, the bifurcated solution is unstable. (See Figure 3.2.)

Chapter 3 – One-Dimensional Stability

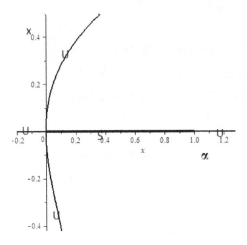

Figure 3.2

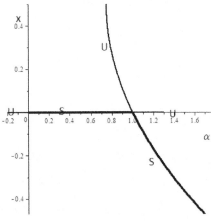

Figure 3.3

Chapter 3 – One-Dimensional Stability

Now consider the solution to Equation 3.12 near (0; 1). This time we let $A(x) = 1 + a_1 x + a_2 x^2 + \cdots$ Substituting this in we get

$$1 + 2a_1 x + (a_1^2 + 2a_1)x^2 + (2a_1 a_2 + 2a_3)x^3 \\ + \cdots - 1 - a_1 x - a_2 x^2 - a_3 x^3 - \cdots \\ + x + a_1 x^2 + a_2 x^3 + \cdots + x^2 = 0.$$

This gives

$$(1 - 1) + (a_1 + 1)x + (a_1^2 + a_2 + a_1 + 1)x^2 \\ + O(x^3) = 0,$$

so $a_1 = -1, a_2 = 1$, and $A(x) = 1 - x + x^2 + O(x^3)$. Where α is less than one, $x=0$ is stable, thus the bifurcated solution is unstable; for $\alpha > 1$ the situation is reversed. (See Figure 3.3.)

Stability is also important for iterative problems. As an example consider the sequence generated by $x_{n+1} = g(x_n; \alpha)$ where $g(x; \alpha) = x^2 - \alpha x + \alpha$. It is easy to check that $x=1$ is a fixed point for all α. Choosing $\alpha = 1.5$ and $x_0 = .9$ we get

$$x_x = .96, x_2 = .9816, x_3 = .99113856, \ldots, x_{10} \\ = .9999331493,$$

Chapter 3 – One-Dimensional Stability

and it certainly appears that x=1 is a stable fixed point. The web diagram is shown in Figure 3.4.

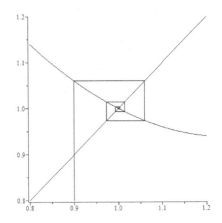

Figure 3.4

Repeating this with $\alpha = 2.5$, we get

$$x_1 = 1.06, x_2 = .9736, x_3 = 1.01389696, x_4 = .993244646, \ldots, x_{10} = .999893529.$$

Again, x=1 appears stable. See Figure 3.5 for the web diagram.

Chapter 3 – One-Dimensional Stability

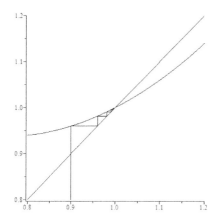

Figure 3.5

Now try this with $\alpha = 0.5$:

$$x_1 = .86, x_2 = .8096, x_3 = .75065216, x_4 = .6881525853, \ldots, x_{10} = .5070085762.$$

This gives us the web diagram in Figure 3.6. The fixed point x=1 is no longer stable. Iterates are approaching x=0.5.

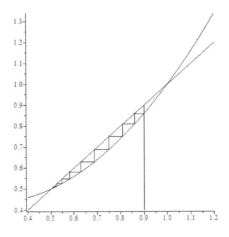

Figure 3.6

So when is a fixed point stable? We want points near the fixed point to get closer, that is, for a given α, if x^* is a fixed point ($g(x^*; \alpha)=x^*$) we want to know when $|g(x; \alpha) - x^*| < |x - x^*|$. Using the fact that $g(x^*; \alpha)=x^*$ and rearranging we get

$$\frac{|g(x; \alpha)-g(x^*; \alpha)|}{|x-x^*|} < 1. \qquad (3.13)$$

Using the Mean Value Theorem we see that this is

$$\left|\frac{\partial g}{\partial x}(\hat{x}; \alpha)\right| < 1 \qquad (3.14)$$

Chapter 3 – One-Dimensional Stability

where $\hat{x}$ is a value between x and x^*. If g is a continuously differentiable function, and $\left|\frac{\partial g}{\partial x}(x^*; \alpha)\right| < 1$, then there is a neighborhood of x^* where Equation 3.14 will hold. If $\left|\frac{\partial g}{\partial x}(x^*; \alpha)\right| > 1$, then there will be a neighborhood where iterates move *further away* from x^*. Putting this together we have the stability criterion for one-dimensional iterative problems:

Stability for One-dimensional Iterative Processes

Let $\{x_n\}$ be a sequence defined by $x_{n+1} = g(x_n; \alpha)$. Suppose that $x^*(\alpha)$ is a fixed point of g ($g(x^*(\alpha); \alpha) = x^*(\alpha)$), and that $g \in C^1$ in a neighborhood of $x^*(\alpha)$. Then

i) If $\left|\frac{\partial g}{\partial x}(x^*(\alpha); \alpha)\right| < 1$, then $x^*(\alpha)$ is stable.

ii) If $\left|\frac{\partial g}{\partial x}(x^*(\alpha); \alpha)\right| > 1$, then $x^*(\alpha)$ is unstable.

iii) If $\left|\frac{\partial g}{\partial x}(x^*(\alpha); \alpha)\right| = 1$, then the test fails.

Chapter 3 – One-Dimensional Stability

Another way of looking at this result is to consider the linear Taylor's expansion. If $x_n = x^* + \varepsilon$, then

$$x_{n+1} = g(x_n; \alpha) = g(x^* + \varepsilon; \alpha)$$
$$= g(x^*; \alpha) + \frac{dg}{dx}(x^*; \alpha)\varepsilon + O(\varepsilon^2)$$
$$= x^* + \frac{dg}{dx}(x^*; \alpha)\varepsilon + O(\varepsilon^2)$$

If ε is sufficiently small and $\left|\frac{\partial g}{\partial x}(x^*; \alpha)\right| < 1$, the iterates are approaching x^*.

Now let us return to our example, $g(x; \alpha) = x^2 - \alpha x + \alpha$. Solving $g(x; \alpha) = x$, we find two fixed points: $x=1$ and $x=\alpha$. $\frac{\partial g}{\partial x} = 2x - \alpha$, so $\frac{\partial g}{\partial x}(1; \alpha) = 2 - \alpha$ and $\frac{\partial g}{\partial x}(x = \alpha; \alpha) = \alpha$. $\frac{\partial g}{\partial x}(1; \alpha) < 1$ for $\alpha > 1$ and $\frac{\partial g}{\partial x}(x = \alpha; \alpha) < 1$ for $\alpha < 1$. There is an exchange of stability at $\alpha = 1$, where the two fixed points cross. (See Figure 3.7.)

Chapter 3 – One-Dimensional Stability

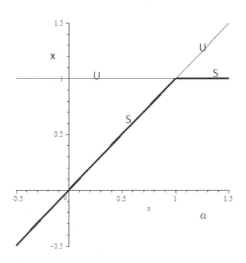

Figure 3.7

If we let $f(x; \alpha) = g(x; \alpha) - x$, then $\frac{\partial f}{\partial x}(x^*(\alpha); \alpha) = \lambda(\alpha) = \frac{\partial g}{\partial x}(x^*(\alpha); \alpha) - 1$. If $\frac{\partial g}{\partial x}(x^*(\alpha); \alpha)$ crosses through the value 1 with non-zero slope, then $\lambda(\alpha)$ crosses through 0 with non-zero slope and our First Bifurcation Theorem applies: there is a unique bifurcated root of f (fixed point of g), other than x^*, in a neighborhood of the point where $\lambda(\alpha)$ crosses zero ($\frac{\partial g}{\partial x}(x^*(\alpha); \alpha)$ crosses through 1).

Further, if g is sufficiently smooth (C^2 locally) the bifurcated solution near the bifurcation point is stable where x^* is unstable and unstable where x^* is stable, just as in our stability theorem for differential equations.

Chapter 3 – One-Dimensional Stability

Chapter 3 Exercises

3.1 For problems 1.2 a & b, find the stability of the steady states. Does this agree with what you found from the numerical solutions?

3.2 For the following, find the steady states, determine their stability, and using a numerical solver and appropriate values of α, demonstrate your results.

   a. $\frac{dx}{dt} = \alpha x + x^3$

   b. $\frac{dx}{dt} = (\alpha - 1 + x^2)(\alpha - x)$

3.3 For the following, find the bifurcation point(s) and expand the bifurcated solution. Predict the stability (using the theorems in this chapter), then use a numerical solver to verify.

   a. $\frac{dx}{dt} = \alpha x - \tan(x)$ (near x=0)

   b. $\frac{dx}{dt} = \alpha x^2 - x^4 - x - \alpha^2 x + \alpha x^3 + \alpha$ (near x= α)

   c. $\frac{dx}{dt} = x(\alpha + \cosh(x))$

Chapter 3 – One-Dimensional Stability

3.4 Get the bifurcation diagram, including stability, for the fixed points of $x_{n+1} = \alpha x_n(1 - x_n)$. Verify numerically. How does this differ from what we found for differential equations?

3.5 Repeat problem 3.4 for $x_{n+1} = x_n^3 + \alpha$. What can you say about the basins of attraction?

3.6 Get the bifurcation and stability diagrams for the following problems. Do these seem to agree with our theorems? Explain.

a. $\frac{dx}{dt} = -x(\alpha^2 - x^2)$

b. $\frac{dx}{dt} = -x(\alpha^2 + x^2)$

c. $\frac{dx}{dt} = -x(\alpha^2 - x)$

Chapter 3 – One-Dimensional Stability

Chapter 4 – Period Doubling

In the last chapter we considered iterative problems, $x_{n+1} = g(x_n; \alpha)$. If $x^*(\alpha)$ is a fixed point ($g(x^*(\alpha); \alpha) = x^*(\alpha)$, then for x near x* we can expand:

$$g(x; \alpha) = x^*(\alpha) + \mu(\alpha)(x - x^*(\alpha)) + (x - x^*(\alpha))h(x - x^*(\alpha); \alpha). \quad (4.1)$$

As we saw in Chapter 3, if $|\mu(\alpha)| < 1$, then the fixed point $x^*(\alpha)$ is stable.

Solving for fixed points of g, we consider $g(x; \alpha)=x$. From Equation 4.1 we get $x^*(\alpha) + \mu(\alpha)(x - x^*(\alpha)) + (x - x^*(\alpha))h(x - x^*(\alpha); \alpha) = x$, or

$$(\mu(\alpha) - 1)(x - x^*(\alpha)) + (x - x^*(\alpha)) \cdot h(x - x^*(\alpha); \alpha) = 0. \quad (4.2)$$

Clearly, $x = x^*(\alpha)$ is a root. Taking the partial with respect to x and setting $x = x^*(\alpha)$, we get $\mu(\alpha) - 1$, so values α_0 such that $\mu(\alpha_0) = 1$ are possible

Chapter 4 - Period Doubling

bifurcation points. As we saw before, if $\frac{d\mu}{d\alpha}(\alpha_0) \neq 0$, then $(x^*(\alpha_0); \alpha_0)$ will be a bifurcation point.

If $\frac{d\mu}{d\alpha}(\alpha_0) \neq 0$, then μ crosses through 1 at α_0. Thus α_0 is a point where the stability of $x=x^*$ changes. As in the continuous case, the bifurcation is an exchange of stability.

However, consider

$$x_{n+1} = x_n^2 - \alpha x_n \qquad (4.3)$$

The fixed points are $x=0$ and $x= \alpha + 1$. For $x=0$ we have $\mu(\alpha) = \left.\frac{\partial g}{\partial x}\right|_{x=0} = -\alpha$, so $x=0$ is stable for $-1 < \alpha < 1$. At $\alpha = -1$, $x=0$ and $x = \alpha + 1$ intersect; $(0; -1)$ is a type-one bifurcation point such as we have been discussing. But what happens at $\alpha = 1$?

Let us chose an α a little larger than 1, say $\alpha = 1.2$. The fixed points of Equation 4.3, $x=0$ and $x=2.2$, are both unstable. Choose an x_0 between the two fixed points, say $x_0 = 1$. You get $x_1 = -.2, x_2 = .28, ..., x_{72} = .558, x_{73} = -.358, ..., x_{200} = .558, x_{201} = -.358, ...$ The sequence appears to settle down to two values that alternate.

Chapter 4 – Period Doubling

Try another α slightly greater than one. Again you will find the sequence converges to an oscillation between two values – a 2-periodic sequence. If we have a sequence generated by $x_{n+1} = g(x_n; \alpha)$, a period-2 point is a point x_1 such that $g(x_1; \alpha) = x_2$ and $g(x_2; \alpha) = x_1$. (Note that if x_1 is a period-2 point, so must $x_2 = g(x_1; \alpha)$ be.)

We can define k-periodic sequences for any positive integer k; a period-a point is the same as a fixed point. Period-k points must appear in multiples of k. Why?

Consider further our period-2 point. We know that $g(x_1; \alpha) = x_2$ and $g(x_2; \alpha) = x_1$. Thus $x_1 = g(x_2; \alpha) = g(g(x_1; \alpha); \alpha) = g^{\circ 2}(x_1; \alpha)$; x_1 is a fixed point of $g^{\circ 2}(x; \alpha)$. For our example we have $g(x; \alpha) = x^2 - \alpha x$. Thus

$$g^{\circ 2}(x; \alpha) = g(g(x; \alpha); \alpha) = g(x^2 - \alpha x; \alpha) = (x^2 - \alpha x)^2 - \alpha(x^2 - \alpha x) \quad (4.4)$$

To find the fixed points of $g^{\circ 2}(x; \alpha)$, we need to solve

$$(x^2 - \alpha x)^2 - \alpha(x^2 - \alpha x) = x^4 - 2\alpha x^3 + (\alpha^2 - \alpha)x^2 + \alpha^2 x = x. \quad (4.5)$$

Chapter 4 – Period Doubling

Factoring a quartic is not normally easy, however, we already know two of the roots. Fixed points of g are also fixed points of $g^{\circ 2}$. Using this we can get

$$x^4 - 2\alpha x^3 + (\alpha^2 - \alpha)x^2 + (\alpha^2 - 1)x = x(x - \alpha - 1)(x^2 + (1 - \alpha)x + 1 - \alpha) \quad (4.6)$$

and we can see that the roots are

$$0, \alpha + 1, \frac{\alpha - 1 \pm \sqrt{\alpha^2 + 2\alpha - 3}}{2}. \quad (4.7)$$

The first two are the fixed points of g. The last two are *not* fixed points of g; they are the actual period-2 points. Plotting them gives us Figure 4.1. At $\alpha = 1$ $g^{\circ 2}$ has a pitchfork bifurcation. It also has one at $\alpha = -3$, which you will consider in the exercises.

Chapter 4 – Period Doubling

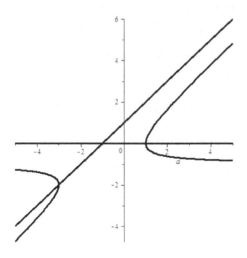

Figure 4.1

What can we say about the stability of these solutions? A 2-periodic sequence is stable if it is stable as a fixed point of $g^{\circ 2}$. We need

$$\frac{\partial}{\partial x} g(g(x;\alpha);\alpha)|_{x=x_1} = \frac{\partial g}{\partial x}(g(x_1;\alpha);\alpha) \cdot \frac{\partial g}{\partial x}(x_1;\alpha) = \frac{\partial g}{\partial x}(x_2;\alpha)\frac{\partial g}{\partial x}(x_1;\alpha). \quad (4.8)$$

Suppose that $x^*(\alpha)$ is a fixed point of g, such that at $\alpha = \alpha_0$, $\frac{\partial g}{\partial x}(x^*(\alpha);\alpha)$ goes through -1. As we have seen, $x^*(\alpha)$ is also a fixed point of $g^{\circ 2}(x;\alpha)$, and $\frac{\partial g^{\circ 2}}{\partial x}(x^*(\alpha);\alpha) = $

Chapter 4 – Period Doubling

$\left(\frac{\partial g}{\partial x}(x^*(\alpha); \alpha)\right)^2$, so $\frac{\partial g^{\circ 2}}{\partial x}(x^*(\alpha); \alpha)$ goes through +1. If $\frac{d\mu}{d\alpha}(\alpha_0) \neq 0$ for x^*, then $\frac{d}{d\alpha}\left(\frac{\partial g^{\circ 2}}{\partial x}(x^*(\alpha); \alpha)\right) \neq 0$ as well; as in our example, $(x^*(\alpha); \alpha)$ is a bifurcation point for $g^{\circ 2}$.

For this example we were able to find the bifurcated 2-periodic solution exactly: the two branches are $\frac{\alpha - 1 \pm \sqrt{\alpha^2 + 2\alpha - 3}}{2}$. In general, we want to be able to get an expansion of this solution. Let us assume that $g(x; \alpha) = \mu(\alpha)x + xh(x; \alpha)$ is analytic, that h is $o(1)$, (actually, since g is analytic, h is $O(x)$), that $\mu(\alpha_0) = -1$, and that $\frac{d\mu}{d\alpha}(\alpha_0) \neq 0$. From what we have already seen. The Implicit Function Theorem guarantees that $x=0$ is the unique fixed point of g in a neighborhood of $(0; \alpha_0)$, and the Implicit Function Theorem for $\frac{1}{x}g^{\circ 2}$ guarantees a bifurcated period-2 solution, x_1 and x_2, such that $g(x_1; \alpha) = x_2$ and $g(x_2; \alpha) = x_1$.

Near the bifurcation point, x_1 and x_2 are both small so we define $\varepsilon = \frac{x_1 - x_2}{2}$ and $y = \frac{x_1 + x_2}{2}$. The Implicit Function Theorem lets us write y as a function of ε, so $x_1 = \varepsilon + y(\epsilon)$ and $x_2 = -\varepsilon + y(\epsilon)$. The way in which we defined ε guarantees

57

Chapter 4 – Period Doubling

that $x_1(-\epsilon) = x_2(\epsilon)$ so y is an even function of ϵ. Similarly, since the bifurcation must be a pitchfork bifurcation, we have that $\alpha = A(x)$ the solution from the Implicit Function Theorem, is the same for x_1 and x_2, thus A is also an even function of ϵ. Putting this all together we have

$$x_1 = \epsilon + \sum_{k=1} b_{2k}\epsilon^{2k}$$
$$x_2 = -\epsilon + \sum_{k=1} b_{2k}\epsilon^{2k} \qquad (4.9)$$
$$\alpha = \alpha_0 + \sum_{k=1} a_{2k}\epsilon^{2k}$$

to be substituted into $g(x_1; \alpha) = x_2$.

For our example we get

$$(\epsilon + b_2\epsilon^2 + b_4\epsilon^4 + \cdots)^2$$
$$- (1 + a_2\epsilon^2 + a_4\epsilon^4 + \cdots)(\epsilon + b_2\epsilon^2 + b_4\epsilon^4 + \cdots)$$
$$= (-\epsilon + b_2\epsilon^4 + b_4\epsilon^4 + \cdots)$$

or

$$(-2b_2 + 1)\epsilon^2 + (-a_2 + 2b_2)\epsilon^3 + \left(-2b_4 - a_2b_2 + b_2{}^2\right)\epsilon^4 + (2b_4 - a_4)\epsilon^5 = O(\epsilon^6). \qquad (4.10)$$

Solving, we get $b_2 = \frac{1}{2}, a_2 = 1, b_4 = \frac{-1}{8}, a_4 = \frac{-1}{4}$, so our expanded solution is

Chapter 4 – Period Doubling

$$x_1 = \varepsilon + \frac{1}{2}\varepsilon^2 - \frac{1}{8}\varepsilon^4 + O(\varepsilon^6)$$
$$x_2 = -\varepsilon + \frac{1}{2}\varepsilon^2 - \frac{1}{8}\varepsilon^4 + O(\varepsilon^6) \quad\quad (4.11)$$
$$\alpha = 1 + \varepsilon^2 - \frac{1}{4}\varepsilon^4 + O(\varepsilon^6)$$

We can compare this to our exact solution, as seen in Figure 4.2.

As α changes, these new fixed points of $g^{\circ 2}$ can also lose stability. If $\frac{\partial g^{\circ 2}}{\partial x}$ at the fixed points of $g^{\circ 2}$ (period-2 points of g) goes through -1, then we will have a period-2 bifurcation of $g^{\circ 2}$, which gives a 4-periodic sequence for g. We can get what is known as a cascade of period doublings, producing periodic points of g of all periods of the form 2^k.

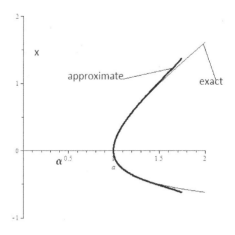

Figure 4.2

Chapter 4 – Period Doubling

It is also possible to have periodic points of other periods, but these come up in a rather different way. We will look at these a bit in a later chapter.

Chapter 4 – Period Doubling

Chapter 4 Exercises

4.1 For the example problem in the chapter, $g(x; \alpha) = x^2 - \alpha x$, expand the bifurcated 2-periodic solutions near $\alpha = -3$. Compare to the exact solution.

4.2 For the problem $g(x; \alpha) = x^2 - \alpha x$ the exact solution is known, so we can check the stability. Where are the period-2 solutions found in the chapter stable? Find where the μ for $g^{\circ 2}$ goes through -1 and expand the period-doubled solutions near there. Numerically generate the sequence near your bifurcation point and compare.

4.3 Analyze $x_{n+1} = \alpha x_n(1 - x_n)$.

4.4 Analyze $x_{n+1} = \dfrac{\alpha}{x_n^2 + 1}$.

Chapter 5 – Other Periods

We have seen how period-doubling leads to periodic points of period 2^k. What about other periods? Certainly we can construct a function that has, for instance, a period-3 point. Let f be a function such that $f(a)=b$, $f(b)=c$, and $f(c)=a$. Plot the three points (a, b), (b, c) and (c, a), then connect them with a continuous function. (See Figure 5.1.)

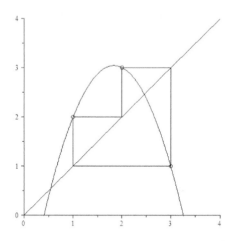

Figure 5.1

For instance, $f(x) = -\frac{3}{2}x^2 + \frac{11}{2}x - 2$ satisfies $f(1)=2$, $f(2)=3$ and $f(3)=1$. Starting at $x_0 = 1$

Chapter 5 – Other Periods

we get a 3-periodic sequence: 1, 2, 3, 1, 2, 3, 1, 2, 3, ... However, if we start at 1.01, we get 2.02485, 2.98665, 1.04646, 2.11292, 2.92442, 1.25597, ... (See Figure 5.2.) What is going on?

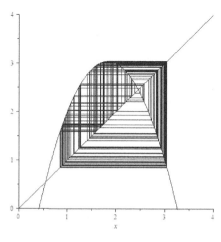

Figure 5.2

A classic result by James Yorke and Tien-Yien Li[1] helps explain this. In 1975 they showed that if a continuous function has a period-3 point, then it also has period-m points for all positive integers m.

---

[1] Li, T.Y. and Yorke, J.A.; *Period Three Implies Chaos*; American Mathematics Monthly; vol.82, no. 10; Dec. 1975; pp. 985-992

Chapter 5 – Other Periods

Yorke & Li's result makes use of the Brower Fixed Point Theorem, which says that a continuous mapping of a connected compact set into, or onto itself has at least one fixed point. This is quite easy to see in one dimension. The graph of a continuous mapping of an interval [u, v] into itself is a continuous curve from the left edge of the square $[u, v] \times [u, v]$ to the right edge. A continuous mapping of [u, v] onto itself is a curve that goes from the top of the square to the bottom. Either must cross the diagonal y=x.

Period Three Implies Chaos

Given a continuous function f, suppose there are values a<b<c, such that $f(a) = b$, $f(b) = c$, and $f(c) \leq a$. (This proof also works, changed as necessary, with a>b>c and $(a) = b, f(b) = c$, and $f(c) \geq a$. For a period-3 point to exist, one or the other must be true.) Define I_1 to be the interval [a, b], and I_2 to be [b, c]. Note that $f(I_1) \supset I_2$ and $f(I_2) \supset (I_1 \cup I_2)$. Since $I_2 \subset f(I_2), f$ has a fixed point. For m>1, define $J_0 = I_1, J_1 = J_2 = \cdots = J_{m-1} = I_2, J_m = I_1$. Now we define a nested sequence of intervals. Let $Q_0 = J_0$. $f(Q_0) \supset J_1$. Define $Q_1 \subset Q_0$

Chapter 5 – Other Periods

such that $f(Q_1) = J_1$. Now $f^{\circ 2}(Q_1) \supset J_2$. Define $Q_2 \subset Q_1$ such that $f^{\circ 2}(Q_2) = J_2$. Continuing this way we can define $Q_k \subset Q_{k-1} \subset \cdots \subset Q_0$ such that $f^{\circ n}(Q_n) = J_n$. Now $f^{\circ m}(Q_m) = J_m = I_1$, and $Q_m \subset Q_0 = I_1$ so $f^{\circ m}$ has a fixed point $x^* \in Q_m$. Since $Q_m \subset Q_{m-1} \subset \cdots \subset Q_0$ we know that $f^{\circ k}(x^*) \in Q_k$ for $k=0,1,2,...,m$. Thus the sequence starting at x^* begins in [a,b], then is in [b,c] for m-1 terms, then goes back to x^* and repeats; it is an m-periodic sequence with no smaller period.

We can actually get more. Given any sequence of positive integers $\{m_j\}$ we can find a starting point such that the sequence generated by f starts in [a, b], spends m_1 iterates in [b, c], then is in [a, b] for one iterate, then in [b, c] for m_2, etc. Define $J_0 = I_1, J_1 = J_2 = \cdots = J_{m_1} = I_2, J_{m_1+1} = I_1, J_{m_1+2} = \cdots = J_{m_1+m_2+1} = I_2$, etc. Define Q_k as before. $Q = \bigcup_k Q_k \neq \emptyset$, (the nested intersection of compact sets is non-empty) and any point $x \in Q$ is such that $f^{\circ j}(x) \in J_j$.

Chapter 5 – Other Periods

The Yorke and Li result is a simple case of a theorem by Sarkovskii. Consider the following ordering of the positive integers:

$$1, 2, 4, 8, \ldots, 2^k, \ldots\ldots\ldots 20, 12, \ldots, 14, 10, 6, \ldots, 9, 7, 5, 3$$
(5.1)

All positive integers appear in the list; a number is either a power of 2, odd, or an odd number times a power of 2. If a function has a periodic point of period m, then it also has periodic points of all periods to the left of m in (5.1). For a nice proof, see Devaney[2].

How do these periodic points appear? The usual way is as a trio of annihilation points. If a, b, c form a 3-cycle under f, then each is a fixed point of $f^{\circ 3}$. To check stability consider $\frac{d}{dx}(f^{\circ 3}(x)) = \frac{d}{dx}(f(f(f(x)))) = \frac{df}{dx}(f(f(x)))\frac{df}{dx}(f(x))\frac{df}{dx}(x)$. At $x=a$ (or b or c) this is $\frac{df}{dx}(a)\frac{df}{dx}(b)\frac{df}{dx}(c)$. Thus the slope of $f^{\circ 3}$ has to be the same at each point of the periodic orbit. Since an annihilation point produces two intersections, one with slope less than one near the bifurcation point and the other greater, we must in fact have a pair of orbits

---

[2] *An Introduction to Chaotic Dynamical Systems*, 2nd edition; Devaney, Robert L.; Addison-Wesley; 1989

Chapter 5 – Other Periods

appear simultaneously, one stable and the other unstable. This is illustrated below for the function $g(x; \alpha) = x^2 - \alpha x$. The graphs are of $g^{\circ 3}(x; \alpha)$. For $\alpha = -1 \pm 2\sqrt{2}$, $g^{\circ 3}(x; \alpha)$ has three double roots. In Figure 5.3 we see $\alpha = -1 + 2\sqrt{2} - .01$, in Figure 5.4 $\alpha = -1 + 2\sqrt{2}$, and in Figure 5.5 $\alpha = -1 + 2\sqrt{2} + .01$. The three humps (not a very technical term) hit the line y=x at the same α value, and cross. The slopes of the intersections match in groups of three.

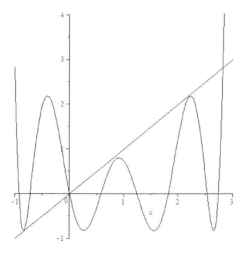

Figure 5.3

Chapter 5 – Other Periods

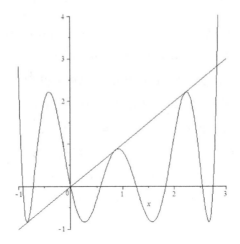

Figure 5.4

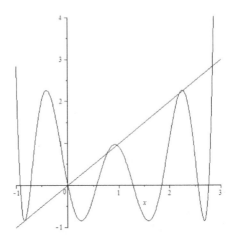

Figure 5.5

When the solutions first appear the slopes of both sets are very close to 1, one just greater and one just smaller. For a neighborhood of α, one

Chapter 5 – Other Periods

of the new periodic points is stable, despite the fact that there are an infinite number of periodic points of other periods between the smallest and largest numbers in our orbit. (See Figure 5.6.)

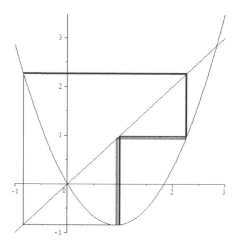

Figure 5.6

The web diagram in Figure 5.6 is for $\alpha = 1.84$. If we try $\alpha = 1.86$, the period-3 solution seems to have lost stability. (See Figure 5.7.)

Chapter 5 – Other Periods

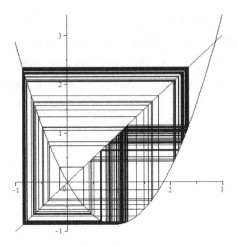

Figure 5.7

The other way that a period-3 point could appear would be for all three points of the orbit to emerge from a fixed point. Again, we would need a pair to keep the parity of the crossing the same. This means that at the bifurcation point g would have to cross $y=x$ with a seventh-order root. This requires $\lambda(\alpha_0) = 1$, and $\frac{d\lambda}{d\alpha}(\alpha_0) = \frac{d\lambda^2}{d\alpha^2}(\alpha_0) = \frac{d\lambda^3}{d\alpha^3}(\alpha_0) = \frac{d\lambda^4}{d\alpha^4}(\alpha_0) = \frac{d\lambda^5}{d\alpha^5}(\alpha_0) = \frac{d\lambda^6}{d\alpha^6}(\alpha_0) = 0$.

Chapter 5 – Other Periods

Chapter 5 Exercises

5.1 Consider the function $g(x; \alpha) = \alpha x(1 - x)$. Find the value(s) of α where period-3 points appear. (There are two ways to do this: solve for where $g^{\circ 3} - x$ has double roots, or for where $\frac{g^{\circ 3} - x}{g - x}$ is equal to a cubic squared.) Plot g and $g^{\circ 3}$ for these values.

5.2 For the same function as in Exercise 5.1, find where the period-3 solutions are stable.

5.3 Numerically, consider $g^{\circ 5}$. Find an approximate interval where g has period-5 points, but not period-3. By Sarkovskii's theorem, g in this interval will have periodic points of all periods other than 3.

5.4 Sarkovskii's theorem only holds in $\mathbb{R}^1$. Consider the mapping in the plane, (in polar coordinates), $(r, \theta) \to \left(\sqrt{r}, \theta + \frac{1}{3}\sin(3\theta) + \frac{2\pi}{3}\right)$. What can you say about periodic points? (First try iterating and see what is going on.)

Chapter 6 – Annihilation Points

There is another important class of bifurcations: annihilation points. These are point where two roots come together and end. (We could have more than two coming together, but this is non-standard. Later we'll see that we cannot have a simple, single root that just ends.) As an example, consider $f(x; \alpha) = x^3 - \alpha x^2 - x^2 + \alpha x - \alpha^2 - \alpha$. Solving $f(x; \alpha) = 0$, we find roots of $x = \alpha, x = \pm\sqrt{-\alpha}$. The two roots $x = \pm\sqrt{-\alpha}$ come together at (0; 0). (See Figure 6.1.)

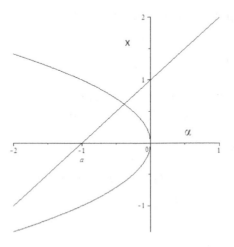

Figure 6.1

Chapter 6 – Annihilation Points

Here we found the annihilation points by being able to find the roots in closed form. In general, this is not possible. What we are seeking is a solution curve $x^*(\alpha)$ such that

$$f(x^*(\alpha); \alpha) = 0, \tag{6.1}$$

and a point $(x^*(\alpha_0); \alpha_0)$ where $\frac{dx^*}{d\alpha}(\alpha_0) = \infty$.
Differentiating Equation 6.1 we get

$$\frac{\partial f}{\partial x}(x^*(\alpha); \alpha) \frac{dx^*}{d\alpha} + \frac{\partial f}{\partial \alpha}(x^*(\alpha); \alpha) = 0. \tag{6.2}$$

Solving for $\frac{dx^*}{d\alpha}$ we have

$$\frac{dx^*}{d\alpha}(\alpha) = \frac{-\frac{\partial f}{\partial \alpha}(x^*(\alpha); \alpha)}{\frac{\partial f}{\partial x}(x^*(\alpha); \alpha)}. \tag{6.3}$$

This will be unboundedly large when

$$\frac{\partial f}{\partial x}(x^*(\alpha); \alpha) = 0 \text{ and } \frac{\partial f}{\partial \alpha}(x^*(\alpha); \alpha) \neq 0.$$

Another way to see this is to think about the Implicit Function Theorem. At an annihilation point, the solution to $f(x; \alpha) = 0$ ceases to be locally unique as a function of α; there are two values of x coming together. Looked at from the side, however, we expect to have $\alpha = A(x)$. Thus we are looking for a point that satisfies

Chapter 6 – Annihilation Points

$$f(x; \alpha) = 0$$
$$\frac{\partial f}{\partial x}(x; \alpha) = 0 \qquad (6.4)$$
$$\frac{\partial f}{\partial \alpha}(x; \alpha) \neq 0$$

For our example, solving $f(x; \alpha) = 0$ and $\frac{\partial f}{\partial x}(x; \alpha) = 0$ we get three points: $(0; 0)$, and $\left(\frac{-1 \pm \sqrt{5}}{2}, \frac{-3 \pm \sqrt{5}}{2}\right)$. Substituting these into $\frac{\partial f}{\partial \alpha}$ we find that the first gives -1 and the other two give 0; (0; 0) is our annihilation point. In a neighborhood of that point, given sufficient smoothness of f, since $\frac{\partial f}{\partial \alpha} \neq 0$, we can expand α as a function of x.

Let us try this for another example. Let $f(x; \alpha) = e^x - \alpha x$. $\frac{\partial f}{\partial x} = e^x - \alpha$. Setting these both equal to 0, we find the point $(1; e)$. Since at this point, $\frac{\partial f}{\partial \alpha} = -x \neq 0$, this is an annihilation point. Set $x = 1 + \varepsilon$ and $\alpha = e + \alpha_2 \varepsilon^2 + \alpha_3 \varepsilon^3 + \cdots$ and substitute into f. (Since $\frac{\partial f}{\partial x} = 0$ we know that $\alpha_1 = 0$.)

$$e^{1+\varepsilon} - (e + \alpha_2 \varepsilon^2 + \alpha_3 \varepsilon^3)(1 + \varepsilon) = 0. \qquad (6.5)$$

Expanding in powers of ε gives

$$\left(-\alpha_2 + \frac{e}{2}\right)\varepsilon^2 + \left(-\alpha_2 - \alpha_3 + \frac{e}{6}\right)\varepsilon^3 + O(\varepsilon^4) = 0,$$

Chapter 6 – Annihilation Points

and solving gives us $\alpha_2 = \frac{e}{2}$ and $\alpha_3 = \frac{-e}{3}$. (Note that we only have an actual annihilation point if the first non-zero term in α is of even order.) Plotting this expansion along with the actual solution gives us Figure 6.2.

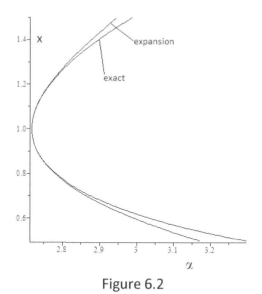

Figure 6.2

The situation is basically the same for roots of $g(x; \alpha) = x$. We look for solutions to

Chapter 6 – Annihilation Points

$$g(x; \alpha) = x$$
$$\frac{\partial g}{\partial x}(x; \alpha) = 1 \qquad (6.6)$$
$$\frac{\partial g}{\partial \alpha}(x; \alpha) \neq 0$$

so that the implicit function theorem will fail in the x direction but not in the α.

Stability is a more interesting question. In the one-dimensional cases we are considering here, we can see that as a pair of roots emerge from a double root, one must have positive slope, and the other negative; one of the two roots that come together is stable and the other is unstable. Similarly, for a pair of fixed points merging, as they cross the line y=x, both must have a slope approaching 1, one of them from above and the other from below; the two fixed points which annihilate each other will have opposite stability. This will turn out not necessarily to be the case in higher dimensions. Two stable points can't come together, but either a stable and an unstable, or two unstable points can.

Chapter 6 – Annihilation Points

Chapter 6 Exercises

6.1   Is Equation 6.4 sufficient to guarantee an annihilation point? Before you answer, consider $f(x; \alpha) = x^3 - \alpha$.

6.2   Find the bifurcation point(s) for $x_{n+1} = \dfrac{\alpha(e^{x_{n-1}} - 1)}{e^{x_{n-1}}}$. Compare your expansion with simulations.

6.3   Find the bifurcation point and expansions of the bifurcated solutions for

$$\frac{dx}{dt} = \tan(x) - \alpha x^3.$$

6.4   Find the bifurcation points and expansions for $\dfrac{dx}{dt} = \alpha - x - \dfrac{5x}{x^4 + 1}$. Sketch the bifurcation diagram, with stability. Compare with some numerical solutions.

6.5   The iterative problem $x_{n+1} = \alpha - \dfrac{5x_n}{x_n^4 + 1}$ has the same roots as the differential equation in Problem 6.4. Is the bifurcation diagram the same? (Hint: not completely.) Find the bifurcation diagram with stability for this problem.

Chapter 7 – Systems of Differential Equations

So far we have been considering single equations, either iterative or differential equations. Many situations, however, lead to systems of equations. Suppose, for instance, we have a population of rabbits whose growth can be described by the logistics equation (per capita growth rate a linear decreasing function of population):

$$\frac{dR}{dT} = R(T)(a - bR(T)) \,. \tag{7.1}$$

Now we add a population of wolves, who eat the rabbits, but die out in the absence of prey. Assuming per capita growth rate of rabbits is now linearly decreasing with W, and the per capita growth rate of W is linearly increasing with R, we get

$$\begin{aligned}\frac{dR}{dT} &= R(a - bR - cW) \\ \frac{dW}{dT} &= W(gR - h)\end{aligned} \tag{7.2}$$

where a, b, c, g and h are constants. Non-dimensionalizing simplifies this a bit to

Chapter 7 – Systems of Differential Equations

$$\frac{dx}{dt} = x(1 - x - y)$$
$$\frac{dy}{dt} = y(\alpha x - \beta)$$
(7.3)

Let us examine this for a couple of different values of α and β. First try $\alpha = 2$ and $\beta = 1$. Several different initial conditions give us Figure 7.1. Using $\alpha = 1$ and $\beta = 2$ gives Figure 7.2.

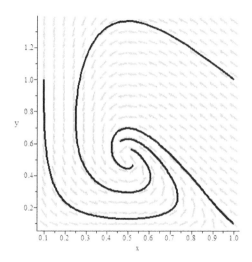

Figure 7.1

Chapter 7 – Systems of Differential Equations

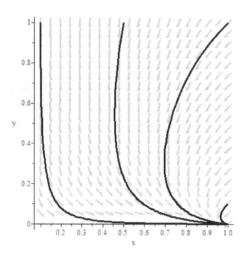

Figure 7.2

As another example, let us consider a single higher-order differential equation. Consider

$$\frac{d^2x}{dt^2} + \frac{dx}{dt} + x(x^2 - \alpha) = 0. \qquad (7.4)$$

By letting $\frac{dx}{dt} = y$ we can write this in system form:

$$\begin{aligned}\frac{dx}{dt} &= y \\ \frac{dy}{dt} &= -y - x(x^2 - \alpha) = 0\end{aligned} \qquad (7.5)$$

Chapter 7 – Systems of Differential Equations

Let us plot solutions of this for a few values of α. With $\alpha = -1$, we get Figure 7.3. Repeating with $\alpha = 1$ gives Figure 7.4.

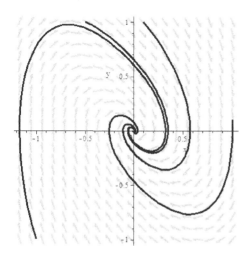

Figure 7.3

Chapter 7 – Systems of Differential Equations

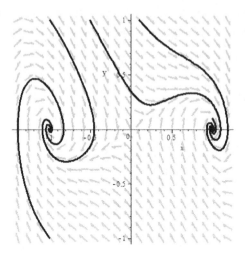

Figure 7.4

Clearly, we have behavior varying with parameters. What can we say in general?

Consider a system of autonomous differential equations

$$\frac{d}{dt}x = f(x;\alpha), f: \mathbb{R}^n \times \mathbb{R} \to \mathbb{R}^n \qquad (7.6)$$

or

$$\frac{dx_i}{dt} = f_i(x_a, \dots, x_n; \alpha), \qquad i = 1, 2, \dots, n.$$

For *n*=2, we can write this as

Chapter 7 – Systems of Differential Equations

$$\frac{dx}{dt} = f(x, y; \alpha)$$
$$\frac{dy}{dt} = g(x, y; \alpha) \qquad (7.7)$$

As before, we look for steady states. A steady state is a vector x_0 in $\mathbb{R}^n$ (possibly depending on α) such that $f(x_0(\alpha); \alpha) = 0$. We want to explore the behavior of solutions near x_0. First consider our special case, Equation 7.7. We solve

$$f(x, y; \alpha) = 0$$
$$g(x, y; \alpha) = 0 \qquad (7.8)$$

And get our steady state $(x_0(\alpha), y_0(\alpha); \alpha)$, a curve in $\mathbb{R}^n \times \mathbb{R}$. Now let

$$x = x_0(\alpha) + \varepsilon x_1(t; \alpha)$$
$$y = y_0(\alpha) + \varepsilon y_1(t; \alpha)$$

Substituting this into Equation 7.7 gives

$$\varepsilon \frac{dx_1}{dt} = f(x_0 + \varepsilon x_1, y_0 + \varepsilon y_1; \alpha)$$
$$\varepsilon \frac{dy_1}{dt} = g(x_0 + \varepsilon x_1, y_0 + \varepsilon y_1; \alpha)$$

Expanding this in a Taylor's expansion in ε, and recalling that (x_0, y_0) satisfies Equation 7.8, we get

Chapter 7 – Systems of Differential Equations

$$\frac{dx_1}{dt} = \frac{\partial f}{\partial x}(x_0, y_0; \alpha)x_1 + \frac{\partial f}{\partial y}(x_0, y_0; \alpha)y_1 + O(\varepsilon)$$

$$\frac{dy_1}{dt} = \frac{\partial g}{\partial x}(x_0, y_0; \alpha)x_1 + \frac{\partial g}{\partial y}(x_0, y_0; \alpha)y_1 + O(\varepsilon)$$

(7.9)

This linearized problem, which in general can be written (letting $x = x_0 + \varepsilon y$) as

$$\frac{d}{dt}\begin{pmatrix} y_1 \\ y_2 \\ \vdots \\ y_n \end{pmatrix} = \begin{pmatrix} \frac{\partial f_1}{\partial y_1} & \frac{\partial f_1}{\partial y_2} & \cdots & \frac{\partial f_1}{\partial y_n} \\ \frac{\partial f_2}{\partial y_1} & \frac{\partial f_2}{\partial y_2} & \cdots & \frac{\partial f_2}{\partial y_n} \\ \vdots & \vdots & \ddots & \vdots \\ \frac{\partial f_n}{\partial y_1} & \frac{\partial f_n}{\partial y_2} & \cdots & \frac{\partial f_n}{\partial y_n} \end{pmatrix}\bigg|_{(x_0;\alpha)} \begin{pmatrix} y_1 \\ y_2 \\ \vdots \\ y_n \end{pmatrix} + O(\varepsilon)$$

(7.10)

is the first two terms of a multivariate Taylor's expansion. The matrix of partial derivatives, often written as $\frac{\partial f}{\partial x}(x_0; \alpha)$, is called the Jacobian. (The absolute value of the determinant of the Jacobian was the scale factor in the change of variables formula for multiple integrals. Unfortunately, the scale factor is also sometimes called the Jacobian.) If solutions to Equation 7.10 approach zero, then trajectories starting sufficiently close to the steady

Chapter 7 – Systems of Differential Equations

state will approach x_0: x_0 is stable. If solutions to 7.10 become unbounded, then x_0 is unstable.

In the next section, we will see how to determine what solutions to 7.10 do.

Chapter 7 – Systems of Differential Equations

Chapter 7 Exercises

7.1 Find the steady states for the two example problems in the book. (Equations 7.3 and 7.5.) Now find the Jacobian at each steady state.

7.2 Find the next term in the expansion given in Equation 7.9.

7.3 Write $\frac{d^2x}{dt^2} + (x^2 + \alpha)\frac{dx}{dt} + x = 0$ as a first order system, and repeat Problem 7.1 for this system.

7.4 Write $\frac{d^2x}{dt^2} + \left(\left(\frac{dx}{dt}\right)^2 + 1\right)\frac{dx}{dt} + x(x - \alpha) = 0$ as a first order system, and repeat Problem 7.1 for this system.

7.5 Find steady states and Jacobians for

$$\frac{dx}{dt} = x(\alpha - xy)$$
$$\frac{dy}{dt} = y(1 - x - y)$$

(Consider only the closed first quadrant.)

Chapter 7 – Systems of Differential Equations

Chapter 8 - Stability of Systems

Near a steady state, the solution to an autonomous system of differential equations behaves like a linear system:

$$\frac{dx}{dt} = Ax. \tag{8.1}$$

If A is a diagonal matrix, we can solve easily:

$$\frac{d}{dt}\begin{pmatrix} x_1 \\ x_2 \\ \vdots \\ x_n \end{pmatrix} = \begin{pmatrix} \lambda_1 & 0 & \cdots & 0 \\ 0 & \lambda_2 & \cdots & 0 \\ \vdots & \vdots & \ddots & \vdots \\ 0 & 0 & \cdots & \lambda_n \end{pmatrix}\begin{pmatrix} x_1 \\ x_2 \\ \vdots \\ x_n \end{pmatrix} \tag{8.2}$$

gives us n uncoupled equations

$$\frac{dx_k}{dt} = \lambda_x x_k, k = 1, \dots, n$$

with solutions

$$x_k(t) = a_k e^{\lambda_k t}. \tag{8.3}$$

If A is not diagonal, but is diagonable, then we can change bases to solve the problem. Suppose $A = PDP^{-1}$ where D is a diagonal matrix.

Chapter 8 – Stability of Systems

Substitute this into Equation 8.1, and multiply through by P^{-1}:

$$P^{-1}\frac{dx}{dt} = \frac{d(P^{-1}x)}{dt} = P^{-1}PDP^{-1}x = D(P^{-1}x).$$
(8.4)

Now we have a solvable system for $y = P^{-1}x$. (We are assuming here that A is a constant matrix.)

This covers almost all constant matrices. If the characteristic polynomial of A has n distinct roots, A is diagonable. The characteristic polynomial and its roots are continuous functions of the entries of the matrix, and a perturbation of a multiple root almost surely splits them.

If we do have a multiple root, A may not be diagonable. In that case, we can define generalized eigenvectors, ψ, based on an eigenvector φ, such that

$$\begin{aligned} A\varphi &= \lambda\varphi \\ A\psi_1 &= \lambda\psi_1 + \varphi \\ &\vdots \\ A\psi_m &= \lambda\psi_m + \psi_{m-1} \end{aligned}$$
(8.5)

Now our equations don't completely uncouple. Take as an example a 2 × 2 case with one generalized eigenvector ($m=1$). Let $P = (\varphi, \psi)$, the matrix whose columns are the eigenvector and the

Chapter 8 – Stability of Systems

generalized eigenvector. Now $PAP^{-1} = J = \begin{pmatrix} \lambda & 1 \\ 0 & \lambda \end{pmatrix}$. Substituting this into Equation 8.1, and again multiplying through by P^{-1}, we get

$$P^{-1}\frac{dx}{dt} = J(P^{-1}x). \qquad (8.6)$$

Let $y = P^{-1}x$, then we have

$$\begin{aligned} \frac{dy_1}{dt} &= \lambda y_1 + y_2 \\ \frac{dy_2}{dt} &= \lambda y_2 \end{aligned} \qquad (8.7)$$

If we solve the second equation first, we can find our solution:

$$\begin{aligned} y_1 &= (a + bt)e^{\lambda t} \\ y_2 &= be^{\lambda t} \end{aligned} \qquad (8.8)$$

In general, if we have more generalized eigenvectors, we find that our solution has the form

$$y_k(t) = (at^{m_k} + bt^{m_k-1} + \cdots + c)e^{\lambda t}. \qquad (8.9)$$

If λ is complex, these formulæ are correct, but not in the form we might want. Let us consider the case where $\lambda = \alpha + i\beta, \beta \neq 0$. If A is real-valued, then $\bar\lambda = \alpha - i\beta$ is also an eigenvalue. Further, if φ is the eigenvector for λ, then $\bar\varphi$ is the eigenvector for $\bar\lambda$. Again, we will first consider the

Chapter 8 – Stability of Systems

2×2 case. Let $\varphi = u + iv$, where u and v are real vectors. Then let Q=(u,v). $Q^{-1}AQ = C = \begin{pmatrix} \alpha & \beta \\ -\beta & \alpha \end{pmatrix}$, so substituting into Equation 8.1 as before, we get

$$\frac{dy_1}{dt} = \alpha y_1 + \beta y_2$$
$$\frac{dy_2}{dt} = -\beta y_1 + \alpha y_2 \qquad (8.10)$$

These are not uncoupled, but solving the first equation algebraically for y_2 and substituting into the second gives

$$\frac{d^2 y_1}{dt^2} - 2\alpha \frac{dy_1}{dt} + (\alpha^2 + \beta^2) y_1 = 0. \qquad (8.11)$$

Solving this, and substituting back for y_2 gives

$$y_1(t) = (a \cdot \cos(\beta t) + b \cdot \sin(\beta t)) e^{\alpha t}$$
$$y_2(t) = (b \cdot \cos(\beta t) - a \cdot \sin(\beta t)) e^{\alpha t} \qquad (8.12)$$

If we had multiple complex roots, we would use the real and imaginary parts of the generalized eigenvectors as well, and have polynomials in t times terms of the form in Equation 8.12.

When we convert our solution back to being in terms of x instead of y, we get linear combinations of functions of the forms we have already seen.

Chapter 8 – Stability of Systems

We are interested in whether or not steady states are stable. If the system in Equation 8.1 came about as the linearization of a nonlinear problem about a steady state (as in Chapter 7), then we want to know whether or not solutions go to zero. Functions of the forms given in 8.3, 8.9 and 8.12 go to zero if $\lambda < 0$, or in the case of complex λ, if $Re(\lambda) = \alpha < 0$. If all the eigenvalues are negative, or have negative real part, then all solutions to 8.1 go to zero: the steady state is stable. If one or more of the eigenvalues have positive real part, then some solutions to 8.1 are unbounded: the steady state is unstable.

There are at least two other ways we can get solutions to Equation 8.1. We can use a new basis directly, or we can find the matrix exponential. First let us solve an example using a new basis.

Consider the problem

$$\frac{d}{dt}\begin{pmatrix}x\\y\end{pmatrix} = \begin{pmatrix}1 & 2\\-4 & -5\end{pmatrix}\begin{pmatrix}x\\y\end{pmatrix}, \quad \begin{pmatrix}x\\y\end{pmatrix}(0) = \begin{pmatrix}3\\1\end{pmatrix}. \quad (8.13)$$

The eigenvalues and eigenvectors of $\begin{pmatrix}1 & 2\\-4 & -5\end{pmatrix}$ are

$$\lambda_1 = -1, \varphi_1 = \begin{pmatrix}1\\-1\end{pmatrix}, \lambda_2 = -3, \varphi_2 = \begin{pmatrix}1\\-2\end{pmatrix}. \quad (8.14)$$

Chapter 8 – Stability of Systems

We will use φ_1 and φ_2 as a basis. Writing everything in terms of the new basis we have

$$\begin{pmatrix} x \\ y \end{pmatrix}(t) = u(t) \begin{pmatrix} 1 \\ -1 \end{pmatrix} + v(t) \begin{pmatrix} 1 \\ -2 \end{pmatrix}$$

$$u(0) \begin{pmatrix} 1 \\ -1 \end{pmatrix} + v(0) \begin{pmatrix} 1 \\ -2 \end{pmatrix} = \begin{pmatrix} 3 \\ 1 \end{pmatrix} = 7 \begin{pmatrix} 1 \\ -1 \end{pmatrix} - 4 \begin{pmatrix} 1 \\ -2 \end{pmatrix}$$

(8.15)

Substituting into Equation 8.13, recalling that the basis vectors are eigenvectors, gives

$$\frac{du}{dt} \begin{pmatrix} 1 \\ -1 \end{pmatrix} + \frac{dv}{dt} \begin{pmatrix} 1 \\ -2 \end{pmatrix} = -u(t) \begin{pmatrix} 1 \\ -1 \end{pmatrix} - 3v(t) \begin{pmatrix} 1 \\ -2 \end{pmatrix}.$$

(8.16)

Using the linear independence of φ_1 and φ_2, Equations 8.15 and 8.16 give

$$\frac{du}{dt} = -u, \quad u(0) = 7$$
$$\frac{dv}{dt} = -3v, \quad v(0) = -4$$

(8.17)

Solving, and substituting back into Equation 8.15 gives

$$\begin{pmatrix} x \\ y \end{pmatrix} = 7e^{-t} \begin{pmatrix} 1 \\ -1 \end{pmatrix} - 4e^{-3t} \begin{pmatrix} 1 \\ -2 \end{pmatrix} =$$
$$\begin{pmatrix} 7e^{-t} - 4e^{-3t} \\ -7e^{-t} + 8e^{-3t} \end{pmatrix}.$$

(8.18)

Chapter 8 – Stability of Systems

This method is equivalent to the first method we discussed, but may be less work for small systems.

Another method of interest is the matrix exponential. Suppose we had the single differential equation

$$\frac{dx}{dt} = ax, \qquad x(0) = x_0.$$

We know the solution: $x(t) = x_0 e^{at}$. Equation 8.1 looks much the same. Can we make sense of the "solution" $x_0 e^{At}$? The first thing to consider is what we mean by e to a matrix power. The function e^{at} can be defined by its Taylor series:

$$e^{at} = 1 + at + \frac{1}{2}(at)^2 + \ldots + \frac{1}{k!}(at)^k + \ldots$$

Let us define e^{At} the same way. To make the addition well-defined, we replace the first 1 with an identity matrix and get

$$e^{At} = I + At + \frac{1}{2}A^2 t^2 + \cdots + \frac{1}{k!}A^k t^k + \cdots \quad (8.19)$$

This raises several questions. First of all, powers of A are only defined for square matrices. Next, consider an example

$$A = \begin{pmatrix} 1 & -2 \\ 1 & 4 \end{pmatrix}.$$

Chapter 8 – Stability of Systems

Raising A to powers gives

$$A^2 = \begin{pmatrix} -1 & -10 \\ 5 & 14 \end{pmatrix}, A^3 = \begin{pmatrix} -11 & -38 \\ 19 & 46 \end{pmatrix}, \dots \quad (8.20)$$

The pattern is not exactly clear.

If A is diagonable we can, again, use $A = PDP^{-1}$. Substituting this into Equation 8.19 gives

$$\begin{aligned} e^{At} &= I + At + \frac{1}{2}A^2t^2 + \cdots + \frac{1}{k!}A^k t^k + \\ &= PP^{-1} + PDP^{-1} + \frac{1}{2}PDP^{-1}PDP^{-1}t^2 + \\ &= PP^{-1} + PDP^{-1} + \frac{1}{2}PD^2P^{-1}t^2 + \\ &= P\left(I + Dt + \frac{1}{2}D^2t^2 + \cdots + \frac{1}{k!}D^k t^k + \cdots\right)P^{-1} \end{aligned}$$

(8.21)

Diagonal matrices are easy to raise to powers – just raise the diagonal elements. This gives

$$e^{At} = P\begin{pmatrix} e^{\lambda_1 t} & 0 & \cdots & 0 \\ 0 & e^{\lambda_2 t} & \cdots & 0 \\ \vdots & \vdots & \ddots & \vdots \\ 0 & 0 & \cdots & e^{\lambda_n t} \end{pmatrix}P^{-1}.$$

(8.22)

Chapter 8 – Stability of Systems

For non-diagonable we can still use the canonical forms. We will have $A = P(D+E)P^{-1}$ where D is a diagonal matrix whose diagonal elements are the real parts of the eigenvalues of A, and D and E commute. This last condition is important since it means that $e^{(D+E)t} = e^{Dt}e^{Et}$. (See the exercises for why.) This then gives

$$e^{At} = Pe^{Dt}e^{Et}P^{-1}. \qquad (8.23)$$

e^{Dt} is the diagonal matrix of the diagonal elements of D, and the elements of e^{Et} are powers of t, sines, cosines, and products of powers of t with sines and cosines. If the eigenvalues of A all have negative real part, e^{At} goes to zero.

Chapter 8 – Stability of Systems

Chapter 8 Exercises

8.1 Find the transforming matrices and put the following in Jordan canonical form.

a. $\begin{pmatrix} 0 & 2 \\ 1 & 1 \end{pmatrix}$

b. $\begin{pmatrix} -3 & 1 \\ -1 & -1 \end{pmatrix}$

c. $\begin{pmatrix} 2 & 1 \\ 1 & 0 \end{pmatrix}$

d. $\begin{pmatrix} -1 & 2 \\ -1 & -3 \end{pmatrix}$

8.2 For the example(s) above with complex eigenvalues, find the transforming matrices and put in real canonical form.

8.3 Using the complex eigenvectors, find the matrix exponential e^{At} for $A = \begin{pmatrix} 0 & \beta \\ -\beta & 0 \end{pmatrix}$. Pick a nonzero vector u and sketch $e^{At}u$ for t increasing. (Think polar.)

8.4 Find the matrix exponentials for the matrices in Problem 8.1.

Chapter 8 – Stability of Systems

8.5 Prove that $\lambda^2 + a\lambda + b = 0$, a and b real, has both roots with negative real part if a and b are both >0.

8.6 Write out the terms in the Taylor expansion of e^A, e^B, and e^{A+B} up to quadratic order. Do they match? What is needed for them to match?

Chapter 8 – Stability of Systems

Chapter 9 – Bifurcation of Systems

Once again, we are faced with the question of what happens when a steady state loses stability. Suppose we have a system

$$\frac{dx}{dt} = f(x; \alpha) \tag{9.1}$$

where $f: \mathbb{R}^n \times \mathbb{R} \to \mathbb{R}^n$. Let us assume $f(0; \alpha) = 0$. (We could shift our variables, as before, if we had a known non-zero solution for all α.) The zero solution to Equation 9.1 is stable if all the eigenvalues of $\frac{\partial f}{\partial x}(0; \alpha)$, the Jacobian evaluated at the steady state, have negative real part. Suppose at $\alpha = \alpha_0$ one of the eigenvalues is zero, i.e., $\lambda(\alpha_0) = 0$, while the others still have negative real part. In the one-dimensional case, when a steady state lost stability, there was a bifurcation. Is that still the case?

First let us consider an example:

$$\frac{d}{dt}\begin{pmatrix} x \\ y \end{pmatrix} = \begin{pmatrix} \alpha x - 2y + \alpha x^2 \\ x + (\alpha - 3)y - xy \end{pmatrix}. \tag{9.2}$$

Chapter 9 – Bifurcation of Systems

We can write this as

$$\frac{d}{dt}\begin{pmatrix}x\\y\end{pmatrix} = \left(\begin{pmatrix}\alpha & -2\\1 & \alpha-3\end{pmatrix} + \begin{pmatrix}\alpha x & 0\\0 & -x\end{pmatrix}\right)\begin{pmatrix}x\\y\end{pmatrix}. \quad (9.3)$$

The Jacobian, $\begin{pmatrix}\alpha & -2\\1 & \alpha-3\end{pmatrix}$, has a characteristic polynomial of $\lambda^2 + (3-2\alpha)\lambda + \alpha^2 - 3\alpha + 2$, and eigenvalues of $(\alpha-1)$ and $(\alpha-2)$. We see that the zero solution is stable for $\alpha < 1$, and loses stability at $\alpha = 1$. Also at $\alpha = 1$ the Implicit Function Theorem fails; the Jacobian is singular. This says that the IFT no longer guarantees a unique solution.

Before, we were able to divide off the known steady state and apply the IFT to what was left. Here, "dividing off" the zero solution leaves us with an operator. The operator, evaluated at (0,0; α), is singular with a zero eigenvector of $\begin{pmatrix}2\\1\end{pmatrix}$. We need to think of how we used the IFT before. We took the derivative not with respect to x, but α. This gave us conditions under which we could solve for α as a function of x, necessarily different from the known solution $x=0$.

Since we expect our bifurcated solution to be close to the zero eigenvector when we are close

Chapter 9 – Bifurcation of Systems

to the bifurcation point, we substitute $\begin{pmatrix} x \\ y \end{pmatrix} = \begin{pmatrix} \varepsilon \\ \varepsilon Y \end{pmatrix}$ into our f (we expect Y to be close to $\frac{1}{2}$). This gives

$$f = \begin{pmatrix} \alpha\varepsilon - 2\varepsilon Y + \alpha\varepsilon^2 \\ \varepsilon + (\alpha - 3)\varepsilon Y - \varepsilon^2 Y \end{pmatrix}$$

or, after canceling an ε,

$$g = \left(\begin{pmatrix} \alpha & -2 \\ 1 & \alpha - 3 \end{pmatrix} + \begin{pmatrix} \alpha\varepsilon & 0 \\ 0 & -\varepsilon \end{pmatrix} \right) \begin{pmatrix} 1 \\ Y \end{pmatrix}. \quad (9.4)$$

We have a point that works: $g\left(\varepsilon = 0, Y = \frac{1}{2}; \alpha = 1\right) = 0$. Taking the Jacobian of g with respect to $(Y; \alpha)$ gives

$$\frac{\partial g}{\partial (Y; \alpha)} = \begin{pmatrix} -2 & 1+\varepsilon \\ \alpha - 3 - \varepsilon & Y \end{pmatrix} \quad (9.5)$$

and at $\left(\varepsilon = 0, Y = \frac{1}{2}; \alpha = 1\right)$ this is non-singular. By the IFT, there is a solution to $g(x,y; \alpha)=0$, given by $(x, Y(x); A(x))$, such that $Y(0)=1/2$ and $A(0)=1$.

If we substitute $x = \varepsilon, Y = \frac{1}{2} + \varepsilon y_1 + \varepsilon^2 y_2 + \cdots, \alpha = 1 + \varepsilon a_1 + \varepsilon^2 a_2 + \cdots$ into Equation 9.4 (or 9.3) and equate powers of ε, we get

$$a_1 = -3, y_1 = -1$$
$$a_2 = 14, y_2 = \frac{11}{2} \quad (9.6)$$
$$\cdots$$

Chapter 9 – Bifurcation of Systems

This gives us an approximate solution of
$\left(\varepsilon, \varepsilon \cdot \left(\frac{1}{2} - \varepsilon + \frac{11}{2}\varepsilon^2 + \cdots\right); 1 - 3\varepsilon + 14\varepsilon^2 + \cdots\right)$.
Plotting this versus the exact solution gives Figure 9.1. (The thicker curve is the exact solution.)

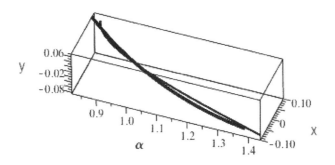

Figure 9.1

In general, suppose we have

$$f(x; \alpha) = (J(\alpha) + H(x; \alpha))x \qquad (9.7)$$

where J is the Jacobian of f at 0, and H is $o(x)$. By assumption, $H(0; \alpha_0) = 0_{n \times n}$, and $J(\alpha_0)$ is singular with $\lambda_1(\alpha_0) = 0$, and all other eigenvalues of J bounded away from the imaginary axis in a neighborhood of α_0. By changing bases if necessary, we can assume that the null vector of $J(\alpha_0)$ is $(1, 0, \cdots, 0)^T$, so

Chapter 9 – Bifurcation of Systems

$$J(\alpha_0) = \begin{pmatrix} \lambda_1(\alpha_0) & 0 & \cdots & 0 \\ 0 & J_{22} & \cdots & J_{2n} \\ \vdots & \vdots & \ddots & \vdots \\ 0 & J_{n2} & \cdots & J_{nn} \end{pmatrix}. \quad (9.8)$$

Now we consider

$$f^*(\varepsilon, y; \alpha) = \frac{1}{\varepsilon} f(\varepsilon, \varepsilon y_2, \ldots, \varepsilon y_n; \alpha) =$$
$$(J(\alpha) + H(\varepsilon(1, y_2, \ldots, y_n); \alpha)) \begin{pmatrix} 1 \\ y \end{pmatrix}. \quad (9.9)$$

We need to take the Jacobian of f^* with respect to $(y; \alpha)$, evaluated at the known zero, ($\varepsilon = 0, y = 0, \alpha = \alpha_0$). since we are not taking derivatives with respect to ε, we can set it equal to zero first. Thus H goes to zero, and

$$\frac{\partial f^*}{\partial (y; \alpha)} = \frac{\partial}{\partial (y; \alpha)} \left(J \begin{pmatrix} 1 \\ y \end{pmatrix} \right) =$$

$$\frac{\partial}{\partial (y; \alpha)} \begin{pmatrix} \lambda_1(\alpha) & 0 & \cdots & 0 \\ 0 & J_{22} & \cdots & J_{2n} \\ \vdots & \vdots & \ddots & \vdots \\ 0 & J_{n2} & \cdots & J_{nn} \end{pmatrix} \begin{pmatrix} 1 \\ y \end{pmatrix} \Bigg|_{\substack{y=0 \\ \alpha=\alpha_0}}$$

$$= \begin{pmatrix} \frac{d\lambda_1}{d\alpha}(\alpha_0) & 0 & \cdots & 0 \\ 0 & J_{22} & \cdots & J_{2n} \\ \vdots & \vdots & \ddots & \vdots \\ 0 & J_{n2} & \cdots & J_{nn} \end{pmatrix}.$$

$$(9.10)$$

Chapter 9 – Bifurcation of Systems

The determinant of this is

$$\frac{d\lambda_1}{d\alpha}(\alpha_0) \prod_{k=2}^{n} \lambda_k(\alpha_0) \qquad (9.11)$$

so if $\frac{d\lambda_1}{d\alpha}(\alpha_0) \neq 0$, and all the other eigenvalues are bounded away from zero, the IFT gives us a bifurcated solution.

It is worth noting that $\frac{d\lambda_1}{d\alpha}(\alpha_0) \neq 0$ if $\frac{d}{d\alpha}(\det(J(\alpha_0))) \neq 0$, since

$$\frac{d}{d\alpha}(\det(J(\alpha_0))) = \frac{d}{d\alpha}(\lambda_1 \prod_{k=2}^{n} \lambda_k) = \frac{d\lambda_1}{d\alpha}\prod_{k=2}^{n}\lambda_k + \lambda_1 \frac{d}{d\alpha}(\prod_{k=2}^{n}\lambda_k) \qquad (9.12)$$

and $\lambda_1(\alpha_0) = 0, \prod_{k=2}^{n} \lambda_k \neq 0$. This is an improvement, since we may not be able to actually find $\lambda_1(\alpha)$. If all eigenvalues of the Jacobian except $\lambda_1(\alpha)$ have negative real part in a neighborhood of α_0, since the eigenvalues of a matrix are continuous functions of the entries, and the entries in the Jacobian are continuous functions of the point at which it is evaluated, all eigenvalues of the linearization around the bifurcated solution except the first have negative real part is a neighborhood of α_0. The sign of the first behaves the same way as in the one-dimensional case. Thus, close to the

Chapter 9 – Bifurcation of Systems

bifurcation point, where x=0 is stable, the bifurcated solution is not, and vice versa.

For another example, consider

$$\frac{dx}{dt} = -x - y - z + x^2$$
$$\frac{dy}{dt} = 5x + y - 3z \qquad (9.13)$$
$$\frac{dz}{dt} = 4x + \alpha y - 4z$$

Clearly, 0 is a solution, and linearizing around it we find the Jacobian

$$\begin{pmatrix} -1 & -1 & -1 \\ 5 & 1 & -3 \\ 4 & \alpha & -4 \end{pmatrix}$$

which has characteristic polynomial

$$\lambda^3 + 4\lambda^2 + (8 + 3\alpha)\lambda + 8\alpha. \qquad (9.14)$$

We can see that there is a zero-eigenvalue at $\alpha = 0$, and that the other two must satisfy $\lambda^2 + 4\lambda + 8 = 0$, so they have negative real part. The α-derivative of the determinant of the Jacobian is -8, so we know we have a bifurcation point.

At $\alpha = 0$, the zero-eigenvector is $\begin{pmatrix} 1 \\ -2 \\ 1 \end{pmatrix}$, so we will look for a solution in the form

Chapter 9 – Bifurcation of Systems

$$x = \varepsilon$$
$$y = -2\varepsilon + y_2\varepsilon^2 + y_3\varepsilon^3 + \cdots$$
$$z = \varepsilon + z_2\varepsilon^2 + z_3\varepsilon^3 + \cdots \quad (9.15)$$
$$\alpha = 0 + \alpha_1\varepsilon + \alpha_2\varepsilon^2 + \cdots$$

Substituting this into the right hand sides of Equation 9.13 and expanding, we find

$$(-y_2 - z_2 + 1)\varepsilon^2 + (-y_3 - z_3)\varepsilon^3 = O(\varepsilon^4)$$
$$(y_2 - 3z_2)\varepsilon^2 + (y_3 - 3z_3)\varepsilon^3 = O(\varepsilon^4)$$
$$(-2\alpha_1 - 4z_2)\varepsilon^2 + (\alpha_1 y_2 - 2\alpha_2 - 4z_3)\varepsilon^3 = O(\varepsilon^4)$$
$$(9.16)$$

Solving, we find $\left\{\alpha_1 = -\frac{1}{2}, y_2 = \frac{3}{4}, z_2 = \frac{1}{4}, \alpha_2 = -\frac{3}{16}, y_3 = 0, z_3 = 0\right\}$. Thus, to order ε^3, we have

$$x = \varepsilon$$
$$y = -2\varepsilon + \frac{3}{4}\varepsilon^2 + 0\varepsilon^3 + \cdots$$
$$z = \varepsilon + \frac{1}{4}\varepsilon^2 + 0\varepsilon^3 + \cdots \quad (9.17)$$
$$\alpha = -\frac{1}{2}\varepsilon - \frac{3}{16}\varepsilon^2 + \cdots$$

Chapter 9 Exercises

9.1 For the following problems, find where the zero solution is stable, find the bifurcation point, and expand the bifurcated solution.

   a. $\frac{d}{dt}\begin{pmatrix}x\\y\end{pmatrix} = \begin{pmatrix}\alpha x + y - xy^2\\x - 2y\end{pmatrix}$

Chapter 9 – Bifurcation of Systems

b. $\frac{d}{dt}\begin{pmatrix} x \\ y \end{pmatrix} + \begin{pmatrix} \alpha y - xy \\ x + y + y^2 \end{pmatrix} = \begin{pmatrix} 0 \\ 0 \end{pmatrix}$

c. $\frac{d}{dt}\begin{pmatrix} x \\ y \end{pmatrix} = \begin{pmatrix} -x + y + xy \\ \alpha x - x - 2x^3 \end{pmatrix}$

9.2 Repeat Problem 9.1 for the following

a. $\frac{d}{dt}\begin{pmatrix} x \\ y \\ z \end{pmatrix} =$

$\begin{pmatrix} x^2 - x - (\alpha + 1)(x + y) \\ x - 6y - 3(z + z^2) \\ y - x \end{pmatrix}$

b. $\frac{d}{dt}\begin{pmatrix} x \\ y \\ z \end{pmatrix} = \begin{pmatrix} ax - 5(x + z) \\ -ax - y - z - zy \\ (\alpha - 1)(y + z) \end{pmatrix}$

9.3 Try to do the same for the following. What goes wrong?

$\frac{d}{dt}\begin{pmatrix} x \\ y \\ z \end{pmatrix} =$

$\begin{pmatrix} \alpha x + (1 - \alpha)y + \alpha x^3 + \alpha^2 x^2 \\ -\alpha x - (\alpha + 1)y + 2z - \alpha x^3 + \alpha^2 x^2 \\ -\alpha y + x^2 \end{pmatrix}$

Chapter 10 – Hopf Bifurcation

We have seen that as an eigenvalue of the Jacobian of a system of differential equations passes through zero, we have an exchange of stability bifurcation. This is not, however, the only possibility in systems. Consider the problem

$$\frac{d}{dt}\begin{pmatrix}x\\y\end{pmatrix} = \begin{pmatrix}y\\-x+(\alpha-x^2)y\end{pmatrix} = \begin{pmatrix}0 & 1\\-1 & \alpha\end{pmatrix}\begin{pmatrix}x\\y\end{pmatrix} + \begin{pmatrix}0\\-x^2 y\end{pmatrix}. \quad (10.1)$$

Setting the right hand equal to zero and solving, we find that the only steady state is $(0, 0;\ \alpha)$. Linearizing around this steady state, we get the Jacobian

$$\begin{pmatrix}0 & 1\\-1 & \alpha\end{pmatrix} \quad (10.2)$$

which has the characteristic polynomial

$$\lambda^2 - \alpha\lambda + 1 = 0. \quad (10.3)$$

Clearly, $\lambda = 0$ is never a root of Equation 10.3. Checking stability, however, we find that for $\alpha < 0$ both roots are either negative or complex with

Chapter 10 – Hopf Bifurcation

negative real part, so the steady state is stable. On the other hand, if $\alpha > 0$, the roots of Equation 10.3 are either two positive numbers, or a complex conjugate pair with positive real part; the steady state is unstable. (Recall Problem 8.4.)

What happens at $\alpha = 0$? There is not a bifurcated steady state. We know that the steady state is unique. Also, the matrix given in Equation 10.2 is never singular, so the IFT does not fail.

At $\alpha = 0$, the Jacobian is $\begin{pmatrix} 0 & 1 \\ -1 & 0 \end{pmatrix}$. This has eigenvalues of $\pm i$. The eigenvalues cross from the negative real part half-plane to the positive real part half-plane as a complex conjugate pair crossing the axis away from 0. If we consider just the linearized equations, the steady state goes from a stable spiral point ($\alpha < 0$) to a center (at $\alpha = 0$), to an unstable spiral point ($\alpha > 0$).

What are solutions to the non-linear problem doing? If we consider $\alpha = 1$, we get

$$\frac{d}{dt}\begin{pmatrix} x \\ y \end{pmatrix} = \begin{pmatrix} y \\ -x + (1 - x^2)y \end{pmatrix}$$

which is equivalent to

$$\frac{d^2 x}{dt^2} + (x^2 - 1)\frac{dx}{dt} + x = 0.$$

Chapter 10 – Hopf Bifurcation

The term multiplying the first derivative acts like the damping coefficient in a mass-spring system. For large x, this equation appears to be damped, but for x less than 1 we have 'negative damping'.

By considering $\frac{dy}{dx} = \frac{-x+(1-x^2)y}{y}$, the slope in the phase plane, we can construct a bounded positively invariant set (a set in the plane that trajectories can enter, but not leave). Since the only steady state is an unstable spiral point, the Poincaré-Bendixon theorem implies the existence of a periodic orbit.

Exploring solutions numerically, we find that for α just larger than 0 we have a small periodic solution, enclosing the steady state, and very close to it. As α grows, so does the periodic orbit. (See Figure 10.1 for α=.005, .01, .02, .1, .2, .5 and 1.) This appearance of a periodic orbit from a steady state is called a **Hopf Bifurcation**.

Chapter 10 – Hopf Bifurcation

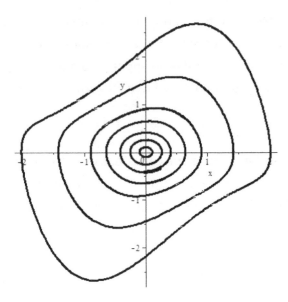

Figure 10.1

In general, suppose we have a system in the plane whose eigenvalues pass through $\pm i\beta_0$. By shifting the origin and changing bases as needed, we can assume our equations are in the form

$$\frac{d}{dt}\begin{pmatrix}x\\y\end{pmatrix} = \left(\begin{pmatrix}\gamma(\alpha) & \beta(\alpha)\\-\beta(\alpha) & \gamma(\alpha)\end{pmatrix} + H(x,y;\alpha)\right)\begin{pmatrix}x\\y\end{pmatrix}$$

(10.4)

where $\gamma(\alpha_0) = 0$ and H is $o(x)$. As you will show in the homework, this can be done without changing the x-axis. Thus, near the steady state, solutions behave like the linearized problem: $\begin{pmatrix}x(t)\\y(t)\end{pmatrix} =$

Chapter 10 – Hopf Bifurcation

$$e^{Jt} \begin{pmatrix} x_0 \\ y_0 \end{pmatrix} = $$
$$e^{\gamma(\alpha)t} \begin{pmatrix} \cos(\beta(\alpha)t) & \sin(\beta(\alpha)t) \\ -\sin(\beta(\alpha)t) & \cos(\beta(\alpha)t) \end{pmatrix} \begin{pmatrix} x_0 \\ y_0 \end{pmatrix}.$$

(10.5)

For $\gamma(\alpha) < 0$ these solutions spiral in; for $\gamma(\alpha) > 0$ they spiral out.

To understand the nonlinear behavior we define the Poincaré map. P is defined on the x-axis sufficiently close to 0. Given a positive x_0, consider the solution to Equation 10.4 starting at $(x = x_0, y = 0)$. Due to the linearized behavior, for x_0 sufficiently small, the trajectory will go around the origin. Now we define $P(x_0) = x_1$, where x_1 is the first intersection of the trajectory with the positive x-axis. We can define P analogously on the negative x-axis, and by continuity, and since 0 is a steady state of Equation 10.4, $P(0)=0$. Any other fixed point of P will represent a periodic trajectory of Equation 10.4. For points close to x=0, the solution is going around the origin in approximately $\frac{2\pi}{\beta}$ time units (the period of the rotation matrix in the linearization). Thus we have

$$P(\varepsilon) = e^{\frac{2\pi}{\beta(\alpha)}\gamma(\alpha)} \varepsilon + O(\varepsilon^2), \qquad (10.6)$$

Chapter 10 – Hopf Bifurcation

assuming that $\beta(\alpha)$ is the positive imaginary part of the complex eigenvalues. Checking stability we find

$$\frac{dP}{dx}(x=0) = e^{\frac{2\pi}{\beta(\alpha)}\gamma(\alpha)} \quad (10.7)$$

and x=0 is stable if $\left|\frac{dP}{dx}(0)\right| < 1$. $\left|e^{\frac{2\pi}{\beta(\alpha)}\gamma(\alpha)}\right| < 1 \leftrightarrow \gamma(\alpha) < 0$; the zero solution is stable as a fixed point of P exactly when it is stable as a steady state of Equation 10.4. Since $e^{\frac{2\pi}{\beta(\alpha)}\gamma(\alpha)}$ is always positive, the fixed point x=0 can only lose stability by having the eigenvalue go through 1. Where this happens, we expect a bifurcated fixed point. To guarantee this we need

$$0 \neq \frac{d}{d\alpha}\left(e^{\frac{2\pi}{\beta(\alpha)}\gamma(\alpha)}\right)\bigg|_{\alpha=\alpha_0} =$$

$$2\pi \frac{\frac{d\gamma}{d\alpha}\beta - \gamma\frac{d\beta}{d\alpha}}{\beta^2} e^{\frac{2\pi}{\beta(\alpha)}\gamma(\alpha)}\bigg|_{\alpha=\alpha_0} = \frac{2\pi}{\beta}\frac{d\gamma}{d\alpha}\bigg|_{\alpha=\alpha_0}. \quad (10.8)$$

(Recall that $\gamma(\alpha_0) = 0$ and $\beta(\alpha_0) \neq 0$.) Thus, if $\frac{d\gamma}{d\alpha}(\alpha_0) \neq 0$, we have a bifurcated fixed point of P, and thus a bifurcated periodic orbit of our original equation.

By the nature of the solution, we must have a pitchfork bifurcation of P. Since the orbits go

Chapter 10 — Hopf Bifurcation

around the origin, a fixed point on the positive x-axis must correspond to another fixed point on the negative axis. This could in general be a stable super-critical pitchfork, or an unstable sub-critical one.

This result in two dimensions goes back at least to Poincaré. The general case, in $\mathbb{R}^n$, is due to Hopf[3]. If a complex conjugate pair of eigenvalues crosses the imaginary axis with non-zero imaginary part and non-zero derivative of the real part, while all other eigenvalues stay in the negative real part half-plane, bounded away from the imaginary axis, there is a bifurcated periodic solution.

(A translation of Hopf's original paper can be found in "The Hopf Bifurcation and its applications"[4].)

Now let us go back to our example, Equation 10.1. We find the eigenvalues (for $|\alpha| < 2$) to be λ and $\bar{\lambda}$, where $\lambda = \frac{\alpha + i\sqrt{4-\alpha^2}}{2}$. For our eigenvectors we choose

---

[3] Hopf, Eberhard; Abzweigung einer periodischen Lösung von einer stationären Lösungn eines Differentialsystems; Sächsichen Akademie der Wissenschaften zu Leipzig; January 1942

[4] Marsden, J. E. and McCracken, M.; The Hopf Bifurcation and its applications; Springer-Verlag; 1976

Chapter 10 – Hopf Bifurcation

$$\varphi = \begin{pmatrix} 1 + i\frac{\alpha}{\sqrt{4-\alpha^2}} \\ i\frac{2}{\sqrt{4-\alpha^2}} \end{pmatrix}. \tag{10.9}$$

Now we use the real and imaginary parts of φ as our basis:

$$\begin{pmatrix} x \\ y \end{pmatrix} = u \begin{pmatrix} 1 \\ 0 \end{pmatrix} + v \begin{pmatrix} \frac{\alpha}{\sqrt{4-\alpha^2}} \\ \frac{2}{\sqrt{4-\alpha^2}} \end{pmatrix}. \tag{10.10}$$

Substituting this into Equation 10.1 gives

$$\frac{d}{dt}\begin{pmatrix} u \\ v \end{pmatrix} = \left(\begin{pmatrix} \frac{\alpha}{2} & \frac{\sqrt{4-\alpha^2}}{2} \\ -\frac{\sqrt{4-\alpha^2}}{2} & \frac{\alpha}{2} \end{pmatrix} + \begin{pmatrix} \frac{2\alpha^2}{4-\alpha^2}v^2 & \frac{\alpha^3}{\sqrt{4-\alpha^2}}v^2 + \frac{\alpha}{\sqrt{4-\alpha^2}}u^2 \\ -\frac{2\alpha}{\sqrt{4-\alpha^2}}v^2 & -\frac{\alpha^2}{4-\alpha^2}v^2 - u^2 \end{pmatrix} \right) \begin{pmatrix} u \\ v \end{pmatrix} \tag{10.11}$$

Now we are looking for a periodic solution, starting at $(u = \varepsilon, v = 0; \alpha)$, circling the origin. (This is the same as starting at $x = \varepsilon, y = 0; \alpha$) since when v=0, x=u and y=v.) We expect u and v to be $O(\varepsilon)$, and since this is a pitchfork bifurcation we expect $\alpha = \alpha_0 + O(\varepsilon^2)$, so we look for our solution in the form

Chapter 10 – Hopf Bifurcation

$$u(t) = \varepsilon u_1(t) + \varepsilon^2 u_2(t) + \cdots$$
$$v(t) = \varepsilon v_1(t) + \varepsilon^2 v_2(t) + \cdots \quad (10.12)$$
$$\alpha = 0 + \varepsilon^2 \alpha_2 + \varepsilon^3 \alpha_3 + \cdots$$

The $O(\varepsilon)$ terms, combined with our assumption about where we start, give us

$$\frac{d}{dt}\begin{pmatrix} u_1 \\ v_1 \end{pmatrix} = \begin{pmatrix} 0 & 1 \\ -1 & 0 \end{pmatrix}\begin{pmatrix} u_1 \\ v_1 \end{pmatrix}, \quad \begin{pmatrix} u_1 \\ v_1 \end{pmatrix}(0) = \begin{pmatrix} 1 \\ 0 \end{pmatrix}. \quad (10.13)$$

The solution to this is

$$\begin{pmatrix} u_1 \\ v_1 \end{pmatrix}(t) = \begin{pmatrix} \cos(t) \\ -\sin(t) \end{pmatrix}. \quad (10.14)$$

The $O(\varepsilon^2)$ terms require

$$\frac{d}{dt}\begin{pmatrix} u_2 \\ v_2 \end{pmatrix} = \begin{pmatrix} 0 & 1 \\ -1 & 0 \end{pmatrix}\begin{pmatrix} u_2 \\ v_2 \end{pmatrix}, \quad \begin{pmatrix} u_2 \\ v_2 \end{pmatrix}(0) = \begin{pmatrix} 0 \\ 0 \end{pmatrix}$$

so

$$\begin{pmatrix} u_2 \\ v_2 \end{pmatrix}(t) = \begin{pmatrix} 0 \\ 0 \end{pmatrix}.$$

Order ε^3 gives

$$\frac{d}{dt}\begin{pmatrix} u_3 \\ v_3 \end{pmatrix} = \begin{pmatrix} 0 & 1 \\ -1 & 0 \end{pmatrix}\begin{pmatrix} u_3 \\ v_3 \end{pmatrix} + \begin{pmatrix} \frac{\alpha_2}{2} & 0 \\ 0 & \frac{\alpha_2}{2} \end{pmatrix}\begin{pmatrix} \cos(t) \\ -\sin(t) \end{pmatrix} + \begin{pmatrix} 0 \\ \cos^2(t)\sin(t) \end{pmatrix}, \quad \begin{pmatrix} u_3 \\ v_3 \end{pmatrix}(0) = \begin{pmatrix} 0 \\ 0 \end{pmatrix}. \quad (10.15)$$

Chapter 10 – Hopf Bifurcation

The solution to this is not bounded unless $\alpha_2 = \frac{1}{4}$. Since we know a periodic solution exists, we know it must be bounded, so $\alpha_2 = \frac{1}{4}$, and $\begin{pmatrix} u_3 \\ v_3 \end{pmatrix} = \begin{pmatrix} \frac{3\sin(t)-\sin(3t)}{32} \\ \frac{3\cos(t)-3\cos(3t)}{32} \end{pmatrix}$. Thus

$$\alpha = \frac{\varepsilon^2}{4} + O(\varepsilon^3), \begin{pmatrix} u \\ v \end{pmatrix} = \varepsilon \begin{pmatrix} \cos(t) \\ -\sin(t) \end{pmatrix} + \varepsilon^3 \begin{pmatrix} \frac{3\sin(t)-\sin(3t)}{32} \\ \frac{3\cos(t)-3\cos(3t)}{32} \end{pmatrix} + O(\varepsilon^4). \qquad (10.16)$$

Now there is good news and bad news. The good news is that we didn't need to work quite so hard. We do not really need the Jacobian to be in canonical form for all α, only at the bifurcation point. This allows us to use a change of basis which is independent of α. The bad news is that we have been assuming that the frequency of the bifurcated solution is constant, which needn't be the case.

Let us try this on another example. Consider

$$\frac{d^2x}{dt^2} + \left(\left(\frac{dx}{dt}\right)^2 + \frac{\alpha-4}{\alpha}\right)\frac{dx}{dt} + \alpha x = 0. \qquad (10.17)$$

Putting this in system form we have

Chapter 10 – Hopf Bifurcation

$$\frac{d}{dt}\begin{pmatrix}x\\y\end{pmatrix} = \begin{pmatrix}y\\-\alpha x - \frac{\alpha-4}{\alpha}y - y^3\end{pmatrix} =$$

$$\begin{pmatrix}0 & 1\\-\alpha & -\frac{\alpha-4}{\alpha}\end{pmatrix}\begin{pmatrix}x\\y\end{pmatrix} + \begin{pmatrix}0\\-y^3\end{pmatrix}. \quad (10.18)$$

The eigenvalues are

$$\lambda = \frac{4-\alpha \pm \sqrt{4\alpha^3 - \alpha^2 + 8\alpha - 16}}{2\alpha} \quad (10.19)$$

and are pure imaginary ($\pm 2i$) at $\alpha = 4$.

At $\alpha = 4$ the Jacobian is

$$\begin{pmatrix}0 & 1\\-4 & 0\end{pmatrix} \quad (10.20)$$

which is not in canonical form. The eigenvector for $\lambda = 2i$ is $\varphi = \begin{pmatrix}1\\2i\end{pmatrix}$, so we use $\left\{\begin{pmatrix}1\\0\end{pmatrix}, \begin{pmatrix}0\\2\end{pmatrix}\right\}$ as our basis. This leads to $x=u$ and $y=2v$, and our systems becomes

$$\frac{d}{dt}\begin{pmatrix}u\\v\end{pmatrix} = \begin{pmatrix}0 & 2\\-\frac{\alpha}{2} & -\frac{\alpha-4}{\alpha}\end{pmatrix}\begin{pmatrix}u\\v\end{pmatrix} + \begin{pmatrix}0\\-4v^3\end{pmatrix} \quad (10.21)$$

which, at $\alpha = 4$, is in canonical form. Again, we choose initial conditions of $\begin{pmatrix}u\\v\end{pmatrix}(0) = \begin{pmatrix}\varepsilon\\0\end{pmatrix}$, since the periodic solution must cross the axis.

Our substitutions are

Chapter 10 – Hopf Bifurcation

$$\tau = (\omega_0 + \varepsilon\omega_1 + \varepsilon^2\omega_2 + \cdots)t$$
$$u(\tau) = \varepsilon u_1(\tau) + \varepsilon^2 u_2(\tau) + \cdots$$
$$v(\tau) = \varepsilon v_1(\tau) + \varepsilon^2 v_2(\tau) + \cdots$$
$$\alpha = 4 + \varepsilon^2 \alpha_2 + \varepsilon^3 \alpha_3 = \cdots$$
(10.22)

The chain rule converts Equation 10.21 to

$$\omega \frac{d}{d\tau}\begin{pmatrix}u\\v\end{pmatrix} = \begin{pmatrix} 0 & 2 \\ -\frac{\alpha}{2} & -\frac{\alpha-4}{\alpha} \end{pmatrix}\begin{pmatrix}u\\v\end{pmatrix} + \begin{pmatrix}0\\-4v^3\end{pmatrix}.$$
(10.23)

(Since we know the frequency of the oscillation at the bifurcation point, we can actually assume that $\omega_0 = 2$.)

Substituting Equation 10.23 into Equation 10.21 and expanding, the $O(\varepsilon)$ terms are

$$\omega_0 \frac{du_1}{d\tau} = 2v_1(\tau)$$
$$\omega_0 \frac{dv_1}{d\tau} = -2u_1(\tau), \quad \begin{pmatrix}u_1\\v_1\end{pmatrix}(0) = \begin{pmatrix}1\\0\end{pmatrix}.$$
(10.24)

With $\omega_0 = 2$ our solutions are

$$\begin{pmatrix}u_1\\v_1\end{pmatrix}(\tau) = \begin{pmatrix}\cos(\tau)\\-\sin(\tau)\end{pmatrix}.$$
(10.25)

The $O(\varepsilon^2)$ terms are

Chapter 10 – Hopf Bifurcation

$$2\frac{du_2}{d\tau} = 2v_2(\tau) + \omega_1\sin(t)$$
$$2\frac{dv_2}{d\tau} = -2u_2(\tau) + \omega_1\cos(t), \quad \begin{pmatrix}u_2\\v_2\end{pmatrix}(0) = \begin{pmatrix}0\\0\end{pmatrix}.$$

(10.26)

The only bounded solution is given by $\omega_1 = 0$, which leads to $\begin{pmatrix}u_2\\v_2\end{pmatrix}(\tau) = \begin{pmatrix}0\\0\end{pmatrix}$.

Continuing with the $O(\varepsilon^3)$ terms, we find that the only periodic solutions come from choosing $\alpha_2 = -12, \omega_2 = -3$. Substituting these in we find

$$\begin{pmatrix}u_3\\v_3\end{pmatrix}(\tau) = \begin{pmatrix} -\frac{3}{16}\sin(\tau) + \frac{1}{16}\sin(3\tau) \\ -\frac{3}{16}\cos(\tau) + \frac{3}{16}\cos(3\tau) + \frac{3}{2}\sin(\tau) \end{pmatrix}.$$

(10.27)

Putting this all together we find

$$\begin{pmatrix}u\\v\end{pmatrix}(t) = \varepsilon \begin{pmatrix} \cos\big((2-3\varepsilon^2)t\big) \\ -\sin\big((2-3\varepsilon^2)t\big) \end{pmatrix}$$

$$+\varepsilon^3 \begin{pmatrix} -\frac{3}{16}\sin\big((2-3\varepsilon^2)t\big) + \frac{1}{16}\sin(3(2-3\varepsilon^2)t) \\ -\frac{3}{16}\cos\big((2-3\varepsilon^2)t\big) + \frac{3}{16}\cos(3(2-3\varepsilon^2)t) + \frac{3}{2}\sin\big((2-3\varepsilon^2)t\big) \end{pmatrix}$$
$$+ O(\varepsilon^4),$$

$$\alpha = 4 - 12\varepsilon^2 + O(\varepsilon^3) \qquad (10.28)$$

More terms can be obtained in the same way.

What we see from all this is the shift in the frequency (2 to $(2 - 3\varepsilon^2)$) and the sine and cosine of 3τ terms breaking the circular symmetry of the periodic orbit. (See Figure 10.2.)

Chapter 10 – Hopf Bifurcation

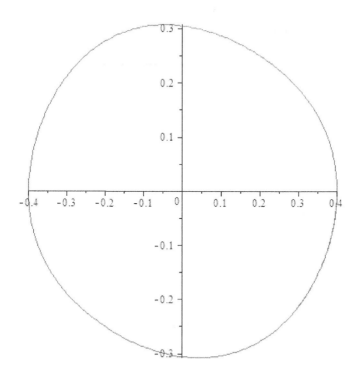

Figure 10.2

Chapter 10 – Hopf Bifurcation

Chapter 10 Exercises

10.1 Carry out the calculations to get Equation 10.27.

10.2 The Jacobian for the example in Equation 10.1 is in canonical form at the bifurcation point so no change of variables is needed. Expand the solution directly, getting two non-zero terms for x and y.

10.3 Find the bifurcated solution for

$$\frac{d^2x}{dt^2} + \left(x^2 + \frac{dx}{dt} - \alpha\right)\frac{dx}{dt} + 9x = 0.$$

10.4 Find the bifurcated solution for

$$\frac{d}{dt}\begin{pmatrix}x\\y\end{pmatrix} = \begin{pmatrix}x+y\\ \alpha y + (\alpha-1)x - x^3 - x^2y\end{pmatrix}.$$

10.5 Analyze

$$\frac{d}{dt}\begin{pmatrix}x\\y\end{pmatrix} = \begin{pmatrix}4x - \alpha y\\ 5x - \alpha y - y^3\end{pmatrix}.$$

10.6 What conditions on the coefficients of $\lambda^2 + a\lambda + b$ are necessary and sufficient for the equation to have pure imaginary roots? What about for $\lambda^3 + a\lambda^2 + b\lambda + c$? (In both cases, assume the coefficients are real numbers.)

Chapter 11 – Sturm Chains

Our results so far have been based on the eigenvalues of the Jacobian of our system. Considering systems in $\mathbb{R}^2$, finding the eigenvalues is straightforward; we can use the quadratic formula. Even if the equation is complicated, we know that the roots of a quadratic both have negative real part if the coefficients all have the same sign. (See Exercise 8.4.) We might guess that this is true for higher order polynomials, but this is not the case. Consider

$$x^3 + x^2 + 3x + 10.$$

The roots are

$$x = -2, \frac{1 \pm i\sqrt{19}}{2}$$

so two of the roots have positive real part.

What can we do in general? For a specific polynomial it is probably easiest to calculate the roots numerically. We, however, are interested in problems with a parameter, which makes numerical solutions difficult. Fortunately there is a result, the Routh-Hurwitz criterion, which will let us count the number of roots in the positive half plane without actually finding them.

To derive the Routh-Hurwitz criterion we must first consider the problem of Sturm – determining the number of distinct real roots of a polynomial in a given real interval. First, define the **Cauchy Index** of a rational function f over an interval $[a, b]$ by

$I[a, b](f)$=(the number of jumps from $-\infty$ to ∞)-(the number of jumps from ∞ to $-\infty$) (11.1)

Each odd-ordered pole in the interval contributes either 1 or -1. What about even-ordered poles? As an example, consider three rational functions:

$$f^- = \frac{1}{(x-2)^2 - \varepsilon^2}$$
$$f = \frac{1}{(x-2)^2}$$
$$f^+ = \frac{1}{(x-2)^2 + \varepsilon^2}$$
(11.2)

Chapter 11 – Sturm Chains

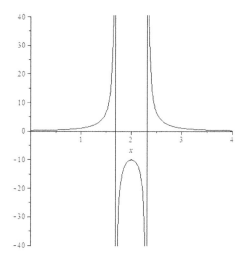

Figure 11.1

The first function, f^-, has two simple poles at $2 \pm \varepsilon$. (See Figure 11.1.) The second, f, has a double pole at 2. (See Figure 11.2.) The third, f^+, has no poles. (See Figure 11.3.) Since f^- has one jump from ∞ to $-\infty$ and another from $-\infty$ to ∞, the Cauchy Index (over any interval containing both poles) is zero. The index of f^+, with no poles, must be zero. Clearly it makes sense for the index of f to also be zero.

Chapter 11 – Sturm Chains

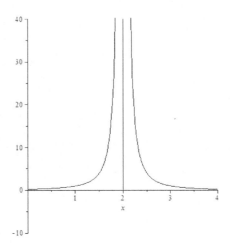

Figure 11.2

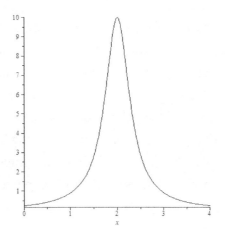

Figure 11.3

Chapter 11 – Sturm Chains

Now suppose our rational function has k poles in the open interval (a, b); we will call them $\{r_i\}_{i=1}^{k}$. Using partial fractions, we can write

$$f(x) = \sum_{i=1}^{k} \sum_{j=1}^{m_i} \left(\frac{a_{ij}}{(x-r_i)^j}\right) + R(x) \quad (11.3)$$

where m_i is the multiplicity of the pole at r_i, the a_{ij}'s are constants, and $R(x)$ is a rational function with no poles in the interval. If m_i is odd and a_{im_i} is positive there is a jump from $-\infty$ to ∞ at r_i; if a_{im_i} is negative there is a jump from ∞ to $-\infty$. (If m_i is even there is no contribution to the index.) Thus

$$I[a,b](f) = \sum_{\substack{i=1 \\ m_i \text{ odd}}}^{k} \operatorname{signum}(a_{im_i}). \quad (11.4)$$

This will be of particular interest for rational functions of the form $\frac{dP/dx}{P}$ where P is a polynomial. In these cases, if $P(x) = a_0 \prod_{i=1}^{k}(x - r_i)^{m_i}$, then

$$\frac{dP/dx}{P} = \sum_{i=1}^{k} \frac{m_i}{x - r_i}. \quad (11.5)$$

Thus $\frac{dP/dx}{P}$ has only simple poles with positive coefficients, and $I[a,b]\left(\frac{dP/dx}{P}\right)$ = the number of distinct real roots of P in the interval (a, b). Once

Chapter 11 – Sturm Chains

we figure out how to evaluate the Cauchy Index of a rational function without factoring, this will let us answer Sturm's problem.

Next, we need the concept of a **Sturm chain**. A Sturm chain of functions on an interval is a set of functions $\{f_1, f_2, \ldots, f_m\}$ such that

(i) For $k = 2,3,\ldots,m-1$ if $f_k(x^*) = 0$ for some x^* in the interval, then $f_{k-1}(x^*) \cdot f_{k+1}(x^*) < 0$

(ii) $f_m(x) \neq 0$ on the interval.

Suppose $\{f_1, f_2, \ldots, f_m\}$ is a Sturm chain. Define $V(x)$ to be the number of sign changes in the sequence $\{f_1(x), f_2(x), \ldots, f_m(x)\}$. If f_k, $k \neq 1$ or m, changes sign at some x_0, then $f_k(x_0) = 0$. This implies that $f_{k-1}(x_0)$ and $f_{k+1}(x_0)$ are of opposite signs. This means that the sequence either went from {...+ + -...} to {...+ - -...} or from {...- + +...} to {...- - +...}. Either way, V stays the same. Since f_m doesn't change sign at all in the interval, we see that V changes only at zeroes of f_1.

If f_1 changes from + to − at x_0 and $f_2(x_0) > 0$, then V goes up by one. If f_1 changes from - to + at x_0 with $f_2(x_0) > 0$, then V goes down by one; vice versa for $f_2(x_0) < 0$. Putting this all together,

Chapter 11 – Sturm Chains

the change in V over an interval is the opposite of the index of $\frac{f_2}{f_1}$, i.e.,

$$I[a,b]\left(\frac{f_2}{f_1}\right) = V(a) - V(b). \quad (11.6)$$

Inspection of Equation 11.6 shows that multiplying the sequence $\{f_1, f_2, \ldots, f_m\}$ by a non-zero function doesn't change Equation 11.6. Thus, our result holds for generalized Sturm chains (Sturm chains multiplied by a non-zero function) as well.

Given two polynomials f_1 and f_2, with $\deg(f_1) \geq \deg(f_2)$, we can construct a Sturm chain by using the division algorithm to define f_k through

$$f_{k-2}(x) = g(x) \cdot f_{k-1}(x) - f_k(x) \quad (11.7)$$

with $\deg(f_k) < \deg(f_{k-1})$. Continue this process until the remainder is 0. If f_1 and f_2 are relatively prime, this will be a Sturm chain. We see from Equation 11.7 that if $f_{k-1}(x_0) = 0$, then $f_{k-2}(x_0) = -f_k(x_0)$; if two adjacent elements of the chain had a common factor it would be a factor of all the elements, contradicting the relative primality of f_1 and f_2. If f_1 and f_2 are not relatively prime, we get a Sturm chain multiplied by their greatest common factor.

Chapter 11 – Sturm Chains

Let us apply this procedure to a polynomial. Let

$$f_1 = P(x) = x^4 + 2x^3 + x^2 - x + 1$$
$$f_2 = \frac{dP}{dx} = 4x^3 + 6x^2 + 2x - 1 \qquad (11.8)$$

Dividing, we see

$$f_1 = \left(\frac{1}{4}x + \frac{1}{8}\right)f_2 + \left(-\frac{1}{4}x^2 - x + \frac{9}{8}\right)$$

so

$$f_3 = \frac{1}{4}x^2 + x - \frac{9}{8}. \qquad (11.9)$$

Continuing, we get

$$f_4 = -60x + 46$$
$$f_5 = \frac{761}{3600} \qquad (11.10)$$

To determine the number of positive real roots, we take $V(0) - V(\infty)$. At x=0, the signs are {+ - - + +}; V(0)=2. At $x = \infty$ the signs are {{+ + + - +} so $V(\infty)$ is also 2: there are no positive real roots.

Repeating this with $P(x) = x^4 + 3x^3 + 2x^2 - x - 1$, we get

Chapter 11 – Sturm Chains

$$f_1 = x^4 + 3x^3 + 2x^2 - x - 1$$
$$f_2 = 4x^3 + 9x^2 + 4x - 1$$
$$f_3 = \frac{11}{16}x^2 + \frac{3}{2}x + \frac{13}{16}$$
$$f_4 = \frac{160}{121}x + \frac{160}{121}$$

(11.11)

and the chain ends, i.e., it is a generalized Sturm chain. $V(-\infty) = 3, V(0) = 1, V(\infty) = 0$, so there are 3-1=2 distinct negative roots, and 1-0=1 distinct positive root. The common factor, f_4, tells us that the negative root at x=-1 is a double one.

Chapter 11 Exercises

11.1 Prove Equation 11.5. (Hint: integrate both sides.)

11.2 Suppose the polynomial P in Equation 11.5 has a pair of complex conjugate roots. How does this affect your conclusion?

11.3 Construct a Sturm chain and use it to count the negative and positive real roots of

$$P(x) = x^5 + 6x^4 + 12x^3 + 14x^2 - 13x - 20.$$

11.4 Repeat Problem 11.3 for

$$P(x) = x^5 + 7x^4 + 16x^3 + 19x^2 + 13x + 4.$$

Chapter 12 – The Routh-Hurwitz Criterion

For questions of stability, we are interested not only in the real eigenvalues, but the real parts of all eigenvalues. Specifically, we want to know the number of roots with positive real part. Let

$$f(z) = a_n z^n + a_{n-1} z^{n-1} + \cdots + a_0 \qquad (12.1)$$

be a polynomial with real coefficients, and assume that f has no roots on the imaginary axis. Consider a semi circle in the right half-plane, centered at the origin, large enough to contain all the roots of f with positive real part. (See Figure 12.1.) If there are, counting multiplicities, k roots of f inside the semicircle,

Chapter 12 – The Routh-Hurwitz Criterion

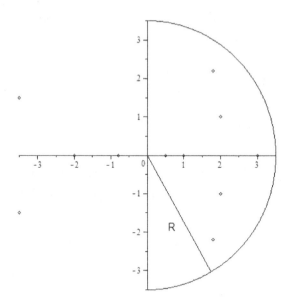

Figure 12.1

then as we traverse the boundary of the semicircular region, the argument of $f(z)$ increases by $2k\pi$. This is true for all semicircles large enough to contain all k roots with positive real part. As the radius, R, becomes large, f is dominated by z^n (assuming $a_n \neq 0$). Thus as we traverse the semicircle from $\arg(z) = -\frac{\pi}{2}$ to $\arg(z) = \frac{\pi}{2}$, the argument of f will increase by $n\pi$. This means that if we can evaluate the change in $\arg(f)$ along the imaginary axis we will be able to solve for k:

$$\left(\arg(f(iy))|_{y=-\infty}^{y=\infty}\right) = (n - 2k)\pi. \quad (12.2)$$

Chapter 12 – The Routh-Hurwitz Criterion

The argument of a complex number is one of the inverse tangents of $\frac{y}{x}$, or one of the inverse cotangents of $\frac{x}{y}$. Let

$$f(iy) = u(y) + iv(y) \qquad (12.3)$$

where u and v are real.

Suppose for now that the degree of u is greater than the degree of v (i.e., n is even). Consider $\arctan\left(\frac{v}{u}(y)\right)$. Since the degree of u is greater than that of v, $\frac{v}{u}$ approaches zero at ∞ and at $-\infty$. Taking $\arctan\left(\frac{v}{u}(-\infty)\right) = 0$, we can evaluate $\arctan\left(\frac{v}{u}(y)\right)$ along the imaginary axis. For example consider

$$f(z) = z^6 - 7z^4 - 7z^3 + 7z + 6. \qquad (12.4)$$

Substituting $z=iy$ and separating real and imaginary parts, we have

$$\begin{aligned} u(y) &= -y^6 - 7y^4 + 6 \\ v(y) &= 7y^3 + 7y \end{aligned} \qquad (12.5)$$

The graph of $\frac{v}{u}(y)$ is shown in Figure 12.2.

Chapter 12 – The Routh-Hurwitz Criterion

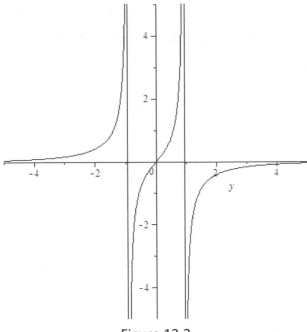

Figure 12.2

As y goes from $-\infty$ to the pole at -1, the $\arctan\left(\frac{v}{u}(y)\right)$ increases from 0 to $\frac{\pi}{2}$. As y continues to increase from -1 to 0, $\arctan\left(\frac{v}{u}(y)\right)$ continues to increase to π, then to $\frac{3\pi}{2}$ at 1, and ends up at 2π as y goes to ∞.

Note that if between two zeroes $\frac{v}{u}$ has a jump from ∞ to $-\infty$, the argument has increased by π. Similarly, if $\frac{v}{u}$ has a jump from $-\infty$ to ∞, the argument decreases by π. Thus we see that

Chapter 12 – The Routh-Hurwitz Criterion

the change in the argument is related to the Cauchy Index by

$$\left(\arg(f(iy))\vert_{y=-\infty}^{y=\infty}\right) = \pi I[-\infty, \infty]\left(\frac{v}{u}\right) \quad (12.6)$$

when deg(v)<deg(u). If deg(v)>deg(u), we find that

$$\left(\arg(f(iy))\vert_{y=-\infty}^{y=\infty}\right) = \pi I[-\infty, \infty]\left(\frac{u}{v}\right). \quad (12.7)$$

Substituting $z=iy$ into Equation 12.1 leads to

$$\left(\arg(f(iy))\vert_{y=-\infty}^{y=\infty}\right) =$$
$$\pi I[-\infty, \infty]\left(\frac{a_{n-1}y^{n-1} - a_{n-3}y^{n-3} + \cdots a_1 y}{a_n y^n - a_{n-2}y^{n-2} + \cdots a_0}\right)$$

$$(12.8)$$

in both cases.

Now we need to construct a Sturm chain with

$$f_1 = a_n y^n - a_{n-2}y^{n-2} + \cdots a_0$$
$$f_2 = a_{n-1}y^{n-1} - a_{n-3}y^{n-3} + \cdots a_1 y \quad (12.9)$$

This is Routh's algorithm[5]. Proceeding as in the last chapter we obtain

---

[5] Routh, E. J.; A treatise on the stability of a given state of motion; London; MacMillan; 1877

Chapter 12 – The Routh-Hurwitz Criterion

$$f_3 = -f_1 + \left(\frac{a_n}{a_{n-1}}y\right) \cdot f_2 = b_{n-2}y^{n-2} - b_{n-4}y^{n-4} + \cdots$$

$$f_4 = -f_2 + \left(\frac{a_{n-1}}{b_{n-2}}y\right) \cdot f_3 = c_{n-3}y^{n-3} - c_{n-5}y^{n-5} + \cdots$$

(12.10)

and so forth. We are first considering the regular case, where none of the leading coefficients are zero. Now we want the change in the number of variations in sign from $-\infty$ to ∞. At $\pm\infty$ the functions are dominated their leading terms, so

$$V(-\infty) - V(\infty) =$$
$$V(a_n, -a_{n-1}, b_{n-2}, -c_{n-3}, \ldots) -$$
$$V(a_n, a_{n-1}, b_{n-2}, c_{n-3}, \ldots). \quad (12.11)$$

It is easily seen that $V(-\infty) = n - V(\infty)$, so $V(-\infty) - V(\infty) = n - 2V(\infty)$. Since the difference in V is $I[-\infty, \infty]\frac{f_2}{f_1}$, which by Equation 12.2 and Equation 12.8 is $n-2k$, we get that

$k=$the number of roots with positive real part$= V(\infty)$. (12.12)

As an example consider

$$f(x) = x^4 + x^3 + 3x^2 + 4x + 6. \quad (12.13)$$

We get

$$f_1 = x^4 - 3x^2 + 6$$
$$f_2 = x^3 - 4x$$

so constructing our Sturm chain gives

$$\begin{aligned} f_1 &= x^4 - 3x^2 + 6 \\ f_2 &= x^3 - 4x \\ f_3 &= xf_2 - f_1 = -x^2 - 6 \\ f_4 &= -xf_3 - f_2 = 10x \\ f_5 &= -\frac{x}{10}f_4 - f_3 = 6 \end{aligned} \qquad (12.14)$$

Thus $V(\infty) = V(1,1,-1,10,6) = 2 = k$, the number of roots with positive real part. Since the roots to Equation 12.13 are $-1 \pm i, \frac{1 \pm i\sqrt{11}}{2}$ we see this is the correct number with positive real part.

This process is possible, but messier, if the coefficients in our polynomial have parameters. For instance, consider the general cubic $ax^3 + bx^2 + cx + d$. Our Sturm chain is

$$\begin{aligned} f_1 &= ax^3 - cx \\ f_2 &= bx^2 - d \\ f_3 &= \frac{ax}{b}f_2 - f_1 = \left(c - \frac{ad}{b}\right)x \\ f_4 &= \frac{b^2 x}{bc-ad}f_3 - f_2 = d \end{aligned} \qquad (12.15)$$

Chapter 12 – The Routh-Hurwitz Criterion

so the number of roots with positive real part is $V\left(a, b, \frac{bc-ad}{b}, d\right)$. In particular, if the leading coefficient (a) is positive, then there will be no roots with positive real part if $b>0$, $d>0$, and $bc>ad$. Higher order polynomials lead to more difficult calculation of the Sturm chain. In 1895, A. Hurwitz developed an easier algorithm[6]. Given a polynomial as in Equation 12.1, construct the $n \times n$ matrix

$$\begin{pmatrix} a_{n-1} & a_{n-3} & a_{n-5} & \cdots & \cdots \\ a_n & a_{n-2} & a_{n-4} & \cdots & \cdots \\ 0 & a_{n-1} & a_{n-3} & a_{n-5} & \cdots \\ 0 & a_n & a_{n-2} & a_{n-4} & \cdots \\ 0 & 0 & a_{n-1} & a_{n-3} & \cdots \\ 0 & 0 & a_n & a_{n-2} & \cdots \\ \cdots & \cdots & \cdots & \cdots & \cdots \end{pmatrix}.$$

(12.16)

The first two rows represent (suppressing minuses) f_1 and f_2 of the Sturm chain. Now subtract $\frac{a_n}{a_{n-1}}$ times the first row from the second, $\frac{a_n}{a_{n-1}}$ times the third row from the fourth, etc. This yields

---

[6] Hurwitz, A.; Ueber die Bedingungen unter welchen eine Gleichung nur Wurzeln mit negative reelen Theilen besizt; Math. Ann.; vol. 46, pp.273-84; 1895

$$\begin{pmatrix} a_{n-1} & a_{n-3} & a_{n-5} & \cdots \\ 0 & b_{n-2} & b_{n-4} & \cdots \\ 0 & a_{n-1} & a_{n-3} & \cdots \\ 0 & 0 & b_{n-2} & \cdots \\ \cdots & \cdots & \cdots & \cdots \end{pmatrix} \quad (12.17)$$

where the second row now represents f_3. Continuing, we subtract $\frac{a_{n-1}}{b_{n-2}}$ times the second, fourth, etc. rows from the third, fifth, etc. rows, and continue to triangulate the matrix:

$$\begin{pmatrix} a_{n-1} & a_{n-3} & a_{n-5} & \cdots & \cdots \\ 0 & b_{n-2} & b_{n-4} & \cdots & \cdots \\ 0 & 0 & c_{n-3} & c_{n-5} & \cdots \\ 0 & 0 & 0 & d_{n-4} & \cdots \\ \cdots & \cdots & \cdots & \cdots & \cdots \end{pmatrix} \quad (12.18)$$

This gives us a matrix with the leading terms of the Sturm chain down the diagonal. If we let D_j be the determinant of the first j rows, first j columns, we can solve for the leading coefficients of the Sturm chain by

$$a_{n-1} = D_1, b_{n-2} = \frac{D_2}{D_1}, c_{n-3} = \frac{D_3}{D_2}, \ldots$$

Thus the number of roots with positive real part is given by

$$V\left(a_n, D_1, \frac{D_2}{D_1}, \dots, \frac{D_n}{D_{n-1}}\right). \qquad (12.19)$$

Notice, however, that the row operations we performed did not affect the determinants of the principle minors, so we needn't have done them. Let H be the matrix in Equation 12.16, and let D_j be the determinant of the principle $j \times j$ minor. Equation 12.19 gives us k, the number of roots with positive real part. In particular, if $a_n > 0$, all roots will have negative real part if and only if all the D_j determinants are positive. This is the Routh-Hurwitz criterion.

As an example consider

$$f(x) = x^4 + ax^3 + 10x^2 + 11x + 9. \qquad (12.20)$$

Constructing our Hurwitz matrix gives

$$H = \begin{pmatrix} a & 11 & 0 & 0 \\ 1 & 10 & 9 & 0 \\ 0 & a & 11 & 0 \\ 0 & 1 & 10 & 9 \end{pmatrix} \qquad (12.21)$$

and our determinants are

$$\begin{aligned} D_1 &= a \\ D_2 &= 10a - 11 \\ D_3 &= -9a^2 + 110a - 121 \\ D_4 &= 9D_3 \end{aligned} \qquad (12.22)$$

Chapter 12 – The Routh-Hurwitz Criterion

The number of roots with positive real part is
$V\left(1, a, \frac{10a-11}{a}, \frac{(11-9a)(a-11)}{10a-11}, 9\right)$, so if $a < 0$, $k = V(+ \ - \ + \ + \ +) = 2$, if $0 < a < \frac{11}{10}$, $k = V(+ \ + \ - \ + \ +) = 2$, and if $\frac{11}{10} < a < \frac{11}{9}$,

$k = V(+ \ + \ + \ - \ +) = 2$. For $\frac{11}{9} < a < 11$, $k = V(+ \ + \ + \ + \ +) = 0$, and for $11 < a$, $k = V(+ \ + \ + \ - \ +) = 2$. Thus we see that for $\frac{11}{9} < a < 11$ there are no positive real part roots; elsewhere there are 2.

Chapter 12 – The Routh-Hurwitz Criterion

Chapter 12 Exercises

12.1 Use the Routh-Hurwitz criterion to determine where $x=0$ is a stable solution to

$$\frac{d^3x}{dt^3} + 2\alpha \frac{d^2x}{dt^2} + \frac{dx}{dt} + (\alpha+1)x - x^2 = 0.$$

12.2 Write the problem in Exercise 12.1 as a first-order system and expand the bifurcated solutions.

12.3 Repeat Problems 12.1 and 12.2 for

$$\frac{d^3x}{dt^3} + 4\frac{d^2x}{dt^2} + (5-\alpha^2-x^3)\frac{dx}{dt} + (2-\alpha)x + \alpha x^3 = 0.$$

12.4 Get the conditions for stability of $x=0$ as a solution of

$$\frac{d^3x}{dt^3} + \alpha \frac{d^2x}{dt^2} + \beta \frac{dx}{dt} + \gamma x = 0.$$

Next, determine what happens at all values where $x=0$ loses stability.

Chapter 13 – Iterative Systems

Just as there are many situations that lead to systems of differential equations, there are also many problems that lead to iterative systems. As a first example, let us consider a bird population model.

A very simple model of a bird population might look for next year's population as a function of this year's:

$$P_{n+1} = f(P_n) \qquad (13.1)$$

The function f needs to be based on things such as survival rates and death rates. If we look at some data, however, we find that for many species the survival rates and birth rates depend on age. In many cases, survival and birth rates are lower the first year, but are basically constant after that. If we make our model with two age classes, chicks (first year) and adults, we will get a second order system.

Suppose 50% of chicks survive the winter and 80% of adults do. Next year's adults will equal .5 times

Chapter 13 – Iterative Systems

this year's chicks plus the surviving .8 of this year's adults:

$$A_{n+1} = .5C_n + .8A_n.$$

If the average number of female eggs hatched per female bird is .1 for the chicks, and .4 for the adults, we get

$$C_{n+1} = .1C_n + .4A_n.$$

Putting this together gives us a linear model

$$\begin{pmatrix} C \\ A \end{pmatrix}_{n+1} = \begin{pmatrix} .1 & .4 \\ .5 & .8 \end{pmatrix} \begin{pmatrix} C \\ A \end{pmatrix}_n. \qquad (13.2)$$

If we start with some small positive number of adults and chicks and iterate, we will find that our population slowly grows; the zero-solution is unstable.

If environmental changes reduce the adult survival rate to 75%, we will have

$$\begin{pmatrix} C \\ A \end{pmatrix}_{n+1} = \begin{pmatrix} .1 & .4 \\ .5 & .75 \end{pmatrix} \begin{pmatrix} C \\ A \end{pmatrix}_n. \qquad (13.3)$$

This time, we will find that our population slowly dwindles; the zero-solution is stable.

What is the crucial difference? As with differential systems, it all depends on the

Chapter 13 – Iterative Systems

eigenvalues. The eigenvalues for the matrix in Equation 13.2 are 1.018 and -0.118; the eigenvalues for Equation 13.3 are 0.978 and -0.128. If the eigenvalues are all less than 1 in absolute value the steady state is stable; if one or more of the eigenvalues is greater than one in absolute value the steady state is unstable.

To see why this is true suppose we have a matrix A with eigenvalues and eigenvectors $\{\lambda_i, \boldsymbol{\varphi}_i\}_{i=1}^{n}$. If these vectors form a basis, than for any initial vector x_0 we can find c_i's such that $x_0 = \sum_{i=1}^{k} c_i \boldsymbol{\varphi}_i$. Then we the define the sequence

$$x_k = A x_{k-1} = A^k x_0 = c_1 \lambda_1^{k} \boldsymbol{\varphi}_1 + c_2 \lambda_2^{k} \boldsymbol{\varphi}_2 + \cdots + c_n \lambda_n^{k} \boldsymbol{\varphi}_n. \tag{13.4}$$

We can see that if all the λ's have magnitude less than one the sequence goes to 0; if any one or more is greater than 1 in magnitude the sequence is unbounded.

The case for non-diagonable matrices is messier, but the result still holds.

Now let us consider nonlinear problems. In our bird model we may want to consider what happens as the population grows. In many bird species, adults are quite territorial and will protect their nesting tree. If there aren't enough nesting trees, the first year survival will go down. Suppose that the first year survival rate is a linearly decreasing function of the adult population:

$$S_1 = \alpha\left(1 - \frac{A}{K}\right).$$

This changes our model to

$$\begin{pmatrix}C\\A\end{pmatrix}_{n+1} = \begin{pmatrix}.1 & .4\\ \alpha & .8\end{pmatrix}\begin{pmatrix}C\\A\end{pmatrix}_n + \begin{pmatrix}0\\ -\frac{A_n C_n}{K}\end{pmatrix}.$$
(13.5)

Near the (0, 0) steady state, the nonlinear terms are negligible, so the stability is determined by the eigenvalues of the matrix in Equation 13.5, which is the Jacobian evaluated at (0, 0). These eigenvalues are

$$\frac{9 \pm \sqrt{49 + 160\alpha}}{20} \qquad (13.6)$$

We can see that when $\alpha = \frac{9}{20}$ the larger eigenvalue hits 1 and the zero solution loses stability.

For this specific example we can easily solve for the fixed points. Setting

$$\begin{pmatrix}C\\A\end{pmatrix} = \begin{pmatrix}.1 & .4\\ \alpha & .8\end{pmatrix}\begin{pmatrix}C\\A\end{pmatrix} + \begin{pmatrix}0\\ -\frac{AC}{K}\end{pmatrix}$$

Chapter 13 – Iterative Systems

and solving we find

$$C = \frac{4}{9}A$$

$$A = 0 \text{ or } \frac{K(20\alpha - 9)}{20}$$

Note that the two solutions, $(C = 0, A = 0)$ and $\left(C = \frac{K(20\alpha-9)}{45}, A = \frac{K(20\alpha-9)}{20}\right)$ cross when $\alpha = \frac{9}{20}$.

In general, if we have a sequence of vectors generated by

$$x_{n+1} = f(x_{n;\,\alpha}) \qquad (13.7)$$

a fixed point is a solution to

$$x = f(x;\,\alpha).$$

A fixed point x^* is stable if all the eigenvalues of the Jacobian, $\frac{\partial f}{\partial x}$, satisfy $|\lambda| < 1$. If at $\alpha = \alpha_0$ one of the eigenvalues is equal to 1, then the linearization of the equation

$$f(x;\,\alpha_0) - x = 0 \qquad (13.8)$$

is singular. (The Jacobian of Equation 13.8 is $\frac{\partial f}{\partial x} - I$.) Thus we can expand the bifurcated fixed point exactly as we did for the steady state in Chapter 9.

As an example consider the system

Chapter 13 – Iterative Systems

$$\begin{pmatrix} x_{n+1} \\ y_{n+1} \end{pmatrix} = \begin{pmatrix} y_n - x_n y_n \\ \alpha x_n + \frac{y_n}{2} - x_n^2 \end{pmatrix}. \qquad (13.9)$$

Linearizing around the zero solution, we find the Jacobian

$$\begin{pmatrix} 0 & 1 \\ \alpha & 1/2 \end{pmatrix}$$

with characteristic polynomial $\lambda^2 - \frac{1}{2}\lambda - \alpha$. We see that at $\alpha = \frac{1}{2}$, $\lambda = 1$ is a root (the other root is $\lambda = -1/2$). Now we are looking for a second solution to

$$\begin{pmatrix} x \\ y \end{pmatrix} = \begin{pmatrix} y - xy \\ \alpha x + \frac{y}{2} - x^2 \end{pmatrix} \qquad (13.10)$$

near $\left(x = 0, y = 0; \alpha = \frac{1}{2}\right)$. Since the eigenvector for $\lambda = 1$ at $\alpha = \frac{1}{2}$ is $\begin{pmatrix} 1 \\ 1 \end{pmatrix}$, we substitute

$$x = \varepsilon$$
$$y = \varepsilon(1 + \varepsilon y_1 + \varepsilon^2 y_2 + \cdots) \qquad (13.11)$$
$$\alpha = \frac{1}{2} + \varepsilon \alpha_1 + \varepsilon^2 \alpha_2 + \cdots$$

into Equation 13.10 and cancel one ε:

Chapter 13 – Iterative Systems

$$\begin{pmatrix} 1 + \varepsilon y_1 + \varepsilon^2 y_2 + \cdots \end{pmatrix} \begin{pmatrix} \dfrac{1}{1 + \varepsilon y_1 + \varepsilon^2 y_2 + \cdots} \end{pmatrix} = \\ \begin{pmatrix} 1 + \varepsilon y_1 + \varepsilon^2 y_2 + \cdots - \varepsilon(1 + \varepsilon y_1 + \varepsilon^2 y_2 + \cdots) \\ \dfrac{1}{2} + \varepsilon \alpha_1 + \varepsilon^2 \alpha_2 + \cdots \dfrac{1 + \varepsilon y_1 + \varepsilon^2 y_2 + \cdots}{2} - \varepsilon \end{pmatrix}$$

(13.12)

The $O(1)$ equation is satisfied. Separating orders gives

$$O(\varepsilon): 0 = y_1 - 1, y_1 = \alpha_1 + \dfrac{y_1}{2} - 1$$
$$O(\varepsilon^2): 0 = y_2 - y_1, y_2 = \alpha_2 + \dfrac{y_2}{2}$$
...

so

$$y_1 = 1, \alpha_1 = \dfrac{3}{2}, y_2 = 1, \alpha_2 = \dfrac{1}{2}, \ldots$$

and our bifurcated solution is

$$\begin{aligned} x &= \varepsilon \\ y &= \varepsilon + \varepsilon^2 + \varepsilon^3 + \cdots \\ \alpha &= \dfrac{1}{2} + \dfrac{3}{2}\varepsilon + \dfrac{1}{2}\varepsilon^2 + \cdots \end{aligned}$$

(13.13)

This agrees with the exact solution, (known in this case) which is

$$y = \dfrac{x}{1-x}, \alpha = \dfrac{1 + 2x - x^2}{2(1-x)}.$$

(13.14)

Chapter 13 – Iterative Systems

Another way that an iterative system can lose stability is through a period-doubling bifurcation. Suppose we have the problem

$$x_{n+1} = f(x_{n;\,\alpha})$$

as in Equation 13.7. If at some $\alpha = \alpha_0$ one of the eigenvalues of $\frac{\partial f}{\partial x}$ crosses through $\lambda = -1$, while the other still satisfy $|\lambda| < 1$, then the fixed point will lose stability there. However, the Jacobian of $f(x;\,\alpha_0) - x$ will still be nonsingular, so the Implicit Function Theorem guarantees a unique solution. Once again we consider fixed points of

$$f^{\circ 2}(x;\,\alpha) = f(f(x;\,\alpha);\,\alpha) \qquad (13.15)$$

The Jacobian of $f^{\circ 2}$ at a fixed point of f is the Jacobian of f, squared. Thus the Jacobian of $f^{\circ 2}$ has an eigenvalue passing through $\lambda = 1$ and thus has a bifurcated fixed point. Since $f^{\circ 2}$ has a bifurcated fixed point but f does not, we have a bifurcated period-two solution.

As an example consider the system

Chapter 13 – Iterative Systems

$$\begin{pmatrix} 0 & \alpha \\ 1 & -1/2 \end{pmatrix} \begin{pmatrix} x_n \\ y_n \end{pmatrix} + \begin{pmatrix} -x_n^2 \\ -x_n y_n \end{pmatrix}. \begin{pmatrix} x_{n+1} \\ y_{n+1} \end{pmatrix} =$$

(13.16)

At $\alpha = \frac{1}{2}$ the linearization about (0,0) has eigenvalues of $\frac{1}{2}$ and -1 and for $\alpha > \frac{1}{2}$, (0,0) is unstable.

Again, we will look for our solutions in the form of solutions to

$$f\big(x(\varepsilon);\ \alpha(\varepsilon^2)\big) = x(-\varepsilon). \qquad (13.17)$$

Since $\begin{pmatrix} 1 \\ -2 \end{pmatrix}$ is the eigenvector for $\lambda = -1$ at $\alpha = \frac{1}{2}$, we can define ε as the average of the first coordinate of the period-2 points and get as our expansions:

$$\begin{aligned} x(\varepsilon) &= \varepsilon + \sum_{k=1} b_{2k} \varepsilon^{2k} \\ y(\varepsilon) &= -2\varepsilon + \sum_{j=2} c_j \varepsilon^j \\ \alpha &= \frac{1}{2} + \sum_{k=1} a_{2k} \varepsilon^{2k} \end{aligned} \qquad (13.18)$$

Chapter 13 – Iterative Systems

Chapter 13 Exercises

13.1 For the following problem find where the linearization about (x=0, y=0) has an eigenvalue of 1. Expand the bifurcated solution nearby and determine stability.

$$\begin{pmatrix} x_{n+1} \\ y_{n+1} \end{pmatrix} = \begin{pmatrix} -\dfrac{\alpha}{2} & 1 - 2\alpha \\ 1 & \dfrac{1}{2} \end{pmatrix} \begin{pmatrix} x_n \\ y_n \end{pmatrix} + \begin{pmatrix} 0 \\ -x_n y_n \end{pmatrix}$$

13.2 Carry out the expansion of the period-2 points in Equation 13.18.

13.3 Consider the non-zero fixed point of Equation 13.5. Where does the Jacobian about this solution have an eigenvalue of -1? Expand the bifurcated period-2 points.

13.4 Compare the expanded solutions in Problem 13.2 to a simulation. Numerically, what happens for larger values of α?

Chapter 14 – Modified Routh-Hurwitz

We saw in Chapter 12 how to use the Routh-Hurwitz criterion to count the number of roots of a polynomial that have positive real part. This let us test for stability of the steady states of a differential system. For iterated systems, on the other hand, we need to count the number of roots with absolute value greater than one.

To answer this we will map the interior of the unit circle in the λ plane to the left half of the z plane. Given λ, let $z = \frac{\lambda-1}{\lambda+1}$:

$$z = \frac{\lambda-1}{\lambda+1} = \frac{\lambda-1}{\lambda+1}\frac{\bar\lambda+1}{\bar\lambda+1} = \frac{|\lambda|^2+\lambda-\bar\lambda-1}{|\lambda+1|^2} = \frac{|\lambda|^2-1}{|\lambda+1|^2} + i\frac{2\,Im(\lambda)}{|\lambda+1|^2}$$

(14.1)

Thus we see that the real part of z will be positive if and only if the absolute value of λ is greater than one.

Suppose we want to know how many roots of

$$f(\lambda) = a_n\lambda^n + a_{n-1}\lambda^{n-1} + \cdots + a_1\lambda + a_0$$

(14.2)

have absolute value greater than one. We solve Equation 14.1 for λ:

$$\lambda = \frac{1+z}{1-z}. \qquad (14.3)$$

Substituting this into Equation 14.2 gives

$$f(\lambda) = f\left(\frac{1+z}{1-z}\right)$$
$$= a_n\left(\frac{1+z}{1-z}\right)^n + a_{n-1}\left(\frac{1+z}{1-z}\right)^{n-1}$$
$$+ \cdots + a_1\left(\frac{1+z}{1-z}\right) + a_0,$$

or, after multiplying through by $(1-z)^n$,

$$g(z) = a_n(1+z)^n + \cdots + a_{n-k}(1+z)^{n-k}(1-z)^k + \cdots + a_1(1+z)(1-z)^{n-1} + a_0(1-z)^n \qquad (14.4)$$

A root of equation 14.2 has absolute value greater than one if the corresponding root of Equation 14.4 has positive real part. Thus applying the Routh-Hurwitz criterion to Equation 14.4 will count the number of roots of Equation 14.2 with $|\lambda| > 1$.

We can tell more. Expanding Equation 14.4 we find

$$g(z) = f(-1)z^n + \cdots + f(1) \qquad (14.5)$$

Chapter 14 – Modified Routh-Hurwitz

so we have $\lambda = 1$ as a root of Equation 14.2 when the constant term of g vanishes, and $\lambda = -1$ is a root when the leading coefficient goes through zero.

Let us put this together in an example. Consider the question of how many roots of

$$\lambda^4 + \frac{3}{2}\lambda^3 + (1+\alpha)\lambda^2 + \frac{1+\alpha}{2}\lambda + \frac{\alpha}{2} \quad (14.6)$$

are outside the unit circle. Using Equation 14.3 gives

$$(1+z)^4 + \frac{3}{2}(1+z)^3(1-z) + (1+\alpha)(1+z)^2(1-z)^2 + \frac{1+\alpha}{2}(1+z)(1-z)^3 + \frac{\alpha}{2}(1-z)^4. \quad (14.7)$$

Expanding gives us

$$\alpha z^4 + (2-\alpha)z^3 + (4+\alpha)z^2 + (6-3\alpha)z + (4+2\alpha). \quad (14.8)$$

The Hurwitz matrix is

$$\begin{pmatrix} 2-\alpha & 6-3\alpha & 0 & 0 \\ \alpha & 4+\alpha & 4+2\alpha & 0 \\ 0 & 2-\alpha & 6-3\alpha & 0 \\ 0 & \alpha & 4+\alpha & 4+2\alpha \end{pmatrix} \quad (14.9)$$

and the determinants are

Chapter 14 – Modified Routh-Hurwitz

$$D_1 = 2 - \alpha, D_2 = 2(2 - \alpha)^2, D_3 = 8(1 - \alpha)(2 - \alpha)^2, D_4 = (4 + 2\alpha)D_3 \qquad (14.10)$$

The number of roots of Equation 14.6 with $|\lambda| > 1$ is the number of roots of Equation 14.8 with $Re(z) > 0$, is

$$V(\alpha, 2 - \alpha, 2(2 - \alpha), 4(1 - \alpha), 4 + 2\alpha) \qquad (14.11)$$

Zeroes occur at $\alpha = -2, 0, 1, 2$ and we find that for

$$\begin{aligned}
\alpha < -2 & \quad k = V(-\ +\ +\ +\ -) = 2 \\
-2 < \alpha < 0 & \quad k = V(-\ +\ +\ +\ +) = 1 \\
0 < \alpha < 1 & \quad k = V(+\ +\ +\ +\ +) = 0 \\
1 < \alpha < 2 & \quad k = V(+\ +\ +\ -\ +) = 2 \\
\alpha > 2 & \quad k = V(+\ -\ -\ +\ +) = 2
\end{aligned} \qquad (14.12)$$

We see that at $\alpha = -2$ a root crosses the unit circle at $\lambda = 1$. At $\alpha = 0$ another root crosses the unit circle, at $\lambda = -1$. A complex conjugate pair $\left(\lambda = \frac{-1 \pm i\sqrt{3}}{2}\right)$ cross the unit circle at $\alpha = 1$. At $\alpha = 1$ we have a zero determinant but no eigenvalues are actually hitting the unit circle. Thus if Equation 14.6 were the characteristic polynomial of the Jacobian about a fixed point, the fixed point would be stable for $0 < \alpha < 1$, and have a period doubling bifurcation at $\alpha = 0$. There would be a

Chapter 14 – Modified Routh-Hurwitz

regular bifurcation from the unstable fixed point at $\alpha = -2$, and there would be a Hopf bifurcation, which we will consider in the next chapter, at $\alpha = 1$.

Chapter 14 – Modified Routh-Hurwitz

Chapter 14 Exercises

14.1 Find necessary and sufficient conditions for the roots of

$$a\lambda^2 + b\lambda + c = 0$$

to satisfy $|\lambda| < 1$.

14.2 Repeat Exercise 14.1 for

$$a\lambda^3 + b\lambda^2 + c\lambda + d = 0.$$

14.3 Suppose

$$\lambda^3 + \alpha\lambda^2 + \frac{1}{2}\lambda - \frac{1}{2}$$

is the characteristic polynomial for the Jacobian at a fixed point. What can you say about stability and bifurcations?

14.4 Repeat Exercise 14.3 for

$$\lambda^4 + \frac{1+\alpha}{2}\lambda^3 + \lambda^2 + \alpha\lambda + \frac{1}{4}.$$

14.5 Find the bifurcation points for

$$\begin{pmatrix} x_{n+1} \\ y_{n+1} \end{pmatrix} = \begin{pmatrix} x_n - y_n - \alpha x_n^2 \\ x_n - \alpha y_n - x_n y_n \end{pmatrix}$$

and expand any regular or period-two bifurcations.

Chapter 14 – Modified Routh-Hurwitz

14.6 Run simulations of the system in Exercise 14.5 near the bifurcation points and compare to your expansions.

Chapter 14 – Modified Routh-Hurwitz

Chapter 15 – Discrete Hopf Bifurcation

We have seen what happens when a fixed point loses stability by having an eigenvalue pass through 1 (bifurcated fixed point) or -1 (period doubling bifurcation). Now let us consider the other possibility, a complex conjugate pair passing through $|\lambda| = 1$.

As an example consider the problem

$$\begin{pmatrix} x_{n+1} \\ y_{n+1} \end{pmatrix} = \begin{pmatrix} \alpha + \cos(\beta) & -\sin(\beta) \\ \sin(\beta) & \alpha + \cos(\beta) \end{pmatrix} \begin{pmatrix} x_n \\ y_n \end{pmatrix} + \begin{pmatrix} -x_n^3 \\ 0 \end{pmatrix} \quad (15.1)$$

with β an angle in the first quadrant. The eigenvalues of the linearization around (0,0) are $\lambda = \alpha + \cos(\beta) \pm i \cdot \sin(\beta) = \alpha + e^{\pm i\beta}$, so we see that (0,0) is stable for $\alpha < 0$ and unstable for $\alpha > 0$.

For α small, near (0,0), we have approximately

Chapter 15 – Discrete Hopf Bifurcation

$$\begin{pmatrix} x_{n+1} \\ y_{n+1} \end{pmatrix} = \begin{pmatrix} \cos(\beta) & -\sin(\beta) \\ \sin(\beta) & \cos(\beta) \end{pmatrix} \begin{pmatrix} x_n \\ y_n \end{pmatrix} \quad (15.2)$$

which is a rotation; $\begin{pmatrix} r \cdot \cos(\theta) \\ r \cdot \sin(\theta) \end{pmatrix}$ is mapped to $\begin{pmatrix} r \cdot \cos(\theta + \beta) \\ r \cdot \sin(\theta + \beta) \end{pmatrix}$. What happens to our iterates? Let us consider a couple of examples.

First taking $\beta = 1$ and $\alpha = -0.01$, and starting at $x_0 = 0.2, y_0 = 0$, we get the picture in Figure 15.1.

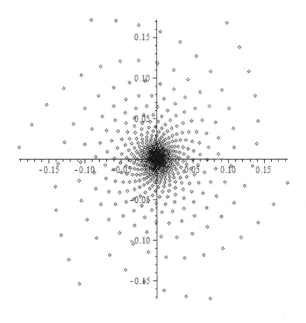

Figure 15.1

Chapter 15 – Discrete Hopf Bifurcation

Next, with the same starting values and β, but with $\alpha = 0.01$, we get Figure 15.2.

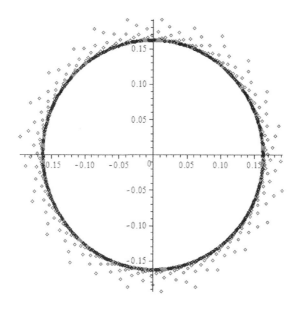

Figure 15.2

Changing our starting position to $x_0 = 0.1, y_0 = 0$ gives Figure 15.3.

Chapter 15 – Discrete Hopf Bifurcation

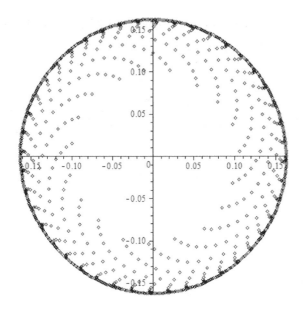

Figure 15.3

It appears that there is some sort of a closed curve to which the iterates converge. To determine the curve in Figures 15.2 and 15.3 it turns out to be easier to work in $\mathbb{C}$, the complex plane. Letting $\begin{pmatrix}x\\y\end{pmatrix} = z, \begin{pmatrix}x\\-y\end{pmatrix} = \bar{z}$, Equation 15.1 becomes

$$z_{n+1} = (\alpha + e^{i\beta})z_n - \frac{1}{8}(z_n + \overline{z_n})^3. \quad (15.3)$$

In general, assuming the linear part of our problem is in canonical form, we get

Chapter 15 – Discrete Hopf Bifurcation

$$z_{n+1} = F(z_n) = \lambda(\alpha)z_n + H(z_n, \overline{z_n}; \alpha). \quad (15.4)$$

By making nonlinear changes of variables, we can eliminate some of the lower-order nonlinearities. First we will eliminate any quadratic terms in H. Assume that

$$H(z_n, \overline{z_n}; \alpha) = h_{11}z_n^2 + h_{12}z_n\overline{z_n} + h_{22}\overline{z_n}^2 + O(z^3). \quad (15.5)$$

Now let

$$u = z + q(z) = z + az^2 + bz\bar{z} + c\bar{z}^2 \quad (15.6)$$

For $z = O(\varepsilon)$ this is locally invertible, giving

$$z = u - au^2 - bu\bar{u} - c\bar{u}^2 + O(\varepsilon^3)$$
$$= u - q(u) + O(\varepsilon^3). \quad (15.7)$$

Now we want to determine our iterative problem for u:

$$u_{n+1} = z_{n+1} + q(z_{n+1}) = F(z_n) + q(F(z_n)) = \\ F(u_n - q(u_n)) + q(F(u_n - q(u_n))) + O(\varepsilon^3) = \\ \lambda(\alpha)u_n - \lambda(\alpha)q(u_n) + H(u_n, \overline{u_n}; \alpha) + \\ q(\lambda(\alpha)u_n) + O(\varepsilon^3).$$

$$(15.8)$$

Using Equations 15.5 and 15.6 we get

Chapter 15 – Discrete Hopf Bifurcation

$$u_{n+1} = \lambda(\alpha)u_n - \lambda(\alpha)\left(au_n^2 + bu_n\overline{u_n} + c\overline{u_n}^2\right) + \left(h_{11}u_n^2 + h_{12}u_n\overline{u_n} + h_{22}\overline{u_n}^2\right) + \left(a\lambda^2 u_n^2 + b\lambda\bar{\lambda}u_n\overline{u_n} + c\bar{\lambda}^2\overline{u_n}^2\right) + O(\varepsilon^3) \quad (15.9)$$

To eliminate the quadratic nonlinearities we solve

$$h_{11} - \lambda a + \lambda^2 = 0$$
$$h_{12} - \lambda b + \lambda\bar{\lambda}b = 0$$
$$h_{22} - \lambda c + \bar{\lambda}^2 c = 0$$

Assuming that $\lambda^k \neq 1, k = 1,2,3$ we can solve these and get

$$a = \frac{h_{11}}{\lambda(\alpha) - \lambda(\alpha)^2}$$
$$b = \frac{h_{12}}{\lambda(\alpha) - \lambda(\alpha)\bar{\lambda}(\alpha)} \quad (15.10)$$
$$c = \frac{h_{22}}{\lambda(\alpha) - \bar{\lambda}(\alpha)^2}$$

These, substituted into Equation 15.6, give us an iterative problem for u,

$$u_{n+1} = \lambda(\alpha)u_n + G(u_n, \overline{u_n}; \alpha) \quad (15.11)$$

where G has only cubic and higher terms.

We can repeat this procedure to eliminate most of the cubic terms. Let

$$w = u + t(u) = u + au^3 + bu^2\bar{u} + cu\bar{u}^2 + d\bar{u}^3. \quad (15.12)$$

Chapter 15 – Discrete Hopf Bifurcation

Assuming $G(u_n, \overline{u_n}; \alpha) = g_{111}u^3 + g_{112}u^2\bar{u} + g_{122}u\bar{u}^2 + g_{222}\bar{u}^3 + O(u^4)$, we get

$$\begin{aligned}
g_{111} - \lambda a + \lambda^3 a &= 0 \\
g_{112} - \lambda b + \lambda^2 \bar{\lambda} b &= 0 \\
g_{122} - \lambda c + \lambda \bar{\lambda}^2 c &= 0 \\
g_{222} - \lambda d + \bar{\lambda}^3 d &= 0
\end{aligned} \quad (15.13)$$

Adding the assumption that $\lambda^4 \neq 1$, we can solve for a, c, and d to eliminate g_{111}, g_{122}, and g_{222}, however, g_{112} is different. We are specifically interested in where $\lambda\bar{\lambda} = 1$, so we can got solve for b and eliminate g_{112}. Setting a, c, and d to the appropriate values, and $b=0$, we get

$$w_{n+1} = \lambda(\alpha)w_n + g_{112}|w_n|^2 w_n + O(w^4) \quad (15.14)$$

Once our problem is in this form, the attracting invariant set we saw in Figures 15.2 and 15.3 is a circle in w-space. Let $\lambda(\alpha) = e^{i\beta} + \lambda_2 \varepsilon^2$ and $w_n = (\varepsilon r_1 + \varepsilon^2 r_2 + \varepsilon^3 r_3 + \cdots)e^{i\theta}$. We want w_{n+1} to have the same radius, so we get

$$w_{n+1} = (\varepsilon r_1 + \varepsilon^2 r_2 + \varepsilon^3 r_3 + \cdots)e^{i\psi} = \lambda(\alpha)w_n + g_{112}|w_n|^2 w_n + O(w^4) = (e^{i\beta} + \lambda_2 \varepsilon^2)(\varepsilon r_1 + \varepsilon^2 r_2 + \varepsilon^3 r_3 + \cdots)e^{i\theta} + g_{112}(\varepsilon r_1 + \varepsilon^2 r_2 + \varepsilon^3 r_3 + \cdots)^3 e^{i\theta} + O(\varepsilon^4).$$

$$(15.15)$$

Chapter 15 – Discrete Hopf Bifurcation

Separating powers of ε we get

$$r_1 e^{i\psi} = r_1 e^{i(\theta+\beta)}$$
$$r_2 e^{i\psi} = r_2 e^{i(\theta+\beta)}$$
$$r_3 e^{i\psi} = r_3 e^{i(\theta+\beta)} + (\lambda_2 r_1 + g_{112} r_1^{\,3}) e^{i\theta}$$
(15.16)

For the circle to be invariant under iterations, we need $\lambda_2 r_1 + g_{112} r_1^{\,3} = 0$, so either $r_1 = 0$ (the fixed point we already know) or

$$r_1 = \sqrt{\frac{\lambda_2}{g_{112}}}. \qquad (15.17)$$

This is the discrete Hopf bifurcation in the plane. Higher dimensional problems can, at least in theory, be reduced to this problem on a stable center manifold. For more details and some discussion of further bifurcations, see Gérard Iooss' book[7].

---

[7] Iooss, G.; Bifurcation of Maps and Applications; North-Holland; 1979

Chapter 15 – Discrete Hopf Bifurcation

Chapter 15 Exercises

15.1 Make the change of variables to simplify the nonlinear terms in the problem given in Equation 15.3.

15.2 Convert the circle of radius $\varepsilon\sqrt{\dfrac{\lambda_2}{g_{112}}}$ in w-space back to z-space.

15.3 Compare your result in Exercise 15.2 to simulations.

Chapter 15 – Discrete Hopf Bifurcation

Chapter 16 - The Brouwer Degree

All of our results so far have been local results. We have expanded solutions near bifurcation points and determined their local stability. Now we would like to consider what happens to these bifurcated solutions.

In 1973 Paul Rabinowitz proved that the solution bifurcating from any characteristic root of odd multiplicity must either be unbounded or contain another characteristic root[8]. His theorem holds in general Banach spaces; we will consider the finite dimensional case.

To prove this result we must first define the Brouwer degree of a mapping. Let $G \subset \mathbb{R}^n$ be a nonempty bounded open set and let $f: \bar{G} \to \mathbb{R}^n$ be continuously differentiable on G and continuous on $\bar{G}$, the closure of G:

$$f \in C^1(G) \cap C(\bar{G}). \tag{16.1}$$

---

[8] Rabinowitz, P.H.; Some aspects of nonlinear eigenvalue problems; Rocky Mountain J. of Mathematics; Vol. 3; No. 2; Spring 1973

Chapter 16 – The Brouwer Degree

Suppose further that

$$f(x) \neq 0, x \in \partial G \qquad (16.2)$$

and that

$\frac{\partial f}{\partial x}$, the Jacobian, is nonsingular at all $x \in G$ such that $f(x) = 0$. $\qquad (16.3)$

Under these assumptions, $f(x) = 0$ has a finite number of solutions in G.

Now we define the degree of f on G as

$$d(f, G) = \sum_{i=1}^{k} signum \left(\det \left(\frac{\partial f}{\partial x}(x_i) \right) \right) \qquad (16.4)$$

where $\{x_i\}_{i=1}^{k}$ are the roots of f in G. If k=0 (f has no roots in G) we define d(f,G)=0.

(We are going to define the Brouwer degree, d(f,G), for $f \in C(\bar{G})$ satisfying Equation 16.2 such that if f also satisfies Equations 16.1 and 16.3 the Brouwer degree will agree with Equation 16.4.)

By defining a family of functions $\varphi_r : \mathbb{R} \to \mathbb{R}$ we can write d(f,G) as an integral. Let φ_r be continuous, with $\varphi_r(0) = 0$; $\varphi_r(t) = 0 \; t \geq r > 0$ and

Chapter 16 – The Brouwer Degree

$$\int_{\mathbb{R}^n} \varphi_r(|x|)\, dx = 1 \qquad (16.5)$$

We will show that for r sufficiently small

$$\int_G \varphi_r(|f(x)|) \det\left(\frac{\partial f}{\partial x}\right) dx = d(f, G) \qquad (16.6)$$

If f has no zeroes in G then $|f(x)|$ is bounded away from zero on $\bar{G}$. Choose r less than this bound and the integrand in Equation 16.6 is identically zero. If f has zeroes at $\{x_i\}_{i=1}^k$ then choose r sufficiently small that

$$f^{-1}(\{z : |z| < r\}) \subset \bigcup_{i=1}^k V_i \qquad (16.7)$$

where the V_is are disjoint open sets in G each containing exactly one x_i, and $f \colon V_j \to f(V_j)$ is one-to-one and onto. This is possible since the Jacobian is nonsingular at all the roots.

Now since

$$\int_{V_i} \varphi_r(|f(x)|) \left|\det\left(\frac{\partial f}{\partial x}\right)\right| dx = \int_{f(V_i)} \varphi_r(|z|)\, dz = \int_{\mathbb{R}^n} \varphi_r(|z|)\, dz = 1 \qquad (16.8)$$

we get that

Chapter 16 – The Brouwer Degree

$$\int_{V_i} \varphi_r(|f(x)|) \det\left(\frac{\partial f}{\partial x}\right) dx =$$
$$signum\left(\det\left(\frac{\partial f}{\partial x}(x_i)\right)\right). \qquad (16.9)$$

Integrating over all of G gives us the sum of the integrals over the $V_i s$ which is the desired result.

Next we need to show that the degree is continuous on functions satisfying conditions 16.1, 16.2 and 16.3. Once we have this we will be able to approximate any $f \in C(\bar{G})$ that satisfies Equation 16.2 by a sequence of $f_j s$ that have degrees defined. Since the degree is continuous on these functions, and integer valued, we can define
$d(f, G) = \lim_{j \to \infty} d(f_j, G)$.

Suppose we have

$$h(x, t): \bar{G} \times [0,1] \to \mathbb{R}^n$$
$$h(x, t) \neq \mathbf{0}, x \in \partial G, t \in [0,1] \qquad (16.10)$$
$$h(x, 0) = f(x), h(x, 1) = g(x)$$

then d(f,G)=d(g,G). This is the **Homotopy Theorem**. As long as we can deform one function into another without introducing any zeroes on the boundary of G, the two functions have the same degree. Essentially, the proof is that we have defined the degree such that $d(h(\cdot, t), G)$ is a continuous function of t. Since $h(x, t) \neq 0, x \in \partial G, t \in [0,1]$, $d(h(\cdot, t), G)$ is defined for all t in $[0,1]$. Continuous integer valued functions are constant on connected components of their domain.

Using the Homotopy Theorem we can prove the following:

> Let $f(x; \alpha)$ be a C^1 function and suppose that $x^*(\alpha)$ is a root of f for $\alpha \in [\alpha_1, \alpha_2]$. Suppose further that $\alpha^*, \alpha_1 < \alpha^* < \alpha_2$, is an isolated root of odd multiplicity of $\det\left(\frac{\partial f}{\partial x}(x^*(\alpha))\right)$. Then $(x^*(\alpha^*); \alpha^*)$ is a bifurcation point.

To prove this let ε be sufficiently small that $\det\left(\frac{\partial f}{\partial x}(x^*(\alpha))\right)$ has no other roots in $[\alpha^* - \varepsilon, \alpha^* + \varepsilon]$. If there is no bifurcation then $\exists r > 0$ such that there are no other roots of f within r of $x^*(\alpha)$. Consider $h(y; \alpha) = f(x^*(\alpha) + y; \alpha)$ and $G = \{y | |y| < r\}$. We can see that

$$d(h(y, \alpha^* - \varepsilon), G) = signum(\det\left(\frac{\partial f}{\partial x}(x^*(\alpha - \varepsilon))\right)$$
and
$$d(h(y, \alpha^* + \varepsilon), G) = signum(\det\left(\frac{\partial f}{\partial x}(x^*(\alpha + \varepsilon))\right),$$

Chapter 16 – The Brouwer Degree

since these are both 'nice' functions with only the one root. By the Homotopy Theorem $d(h(y, \alpha^* - \varepsilon), G) = d(h(y, \alpha^* + \varepsilon), G)$, but since α^* is a root of odd multiplicity there must have been a change of sign. This is a contradiction, so there must in fact have been a bifurcation.

There are several other useful properties of the Brouwer degree:

The Additive Property

If $\bar{G} = \bigcup_{i=1}^{k} \bar{G}_i$ where the $G_i s$ are disjoint open sets, and $f(x) \neq 0, x \in \partial G_i, i = 1, 2, \dots, k$ then $d(f, G) = \sum_{i=1}^{k} d(f, G_i)$.

The Excision Property

If $\bar{K} \subset \bar{G}$ and $f(x) \neq 0$ on $\bar{K}$, then $d(f, G) = d(f, G - K)$.

Finally, let G be an open set in $\mathbb{R}^n \times \mathbb{R}$ and define $G_{[a]}$ to be the 'slice' at a, that is $G_{[a]} + \{(x; \alpha) \in G | \alpha = a\}$. We can now give

The Set Homotopy Theorem

Let G be an open set in $\mathbb{R}^n \times \mathbb{R}$ and consider $f(x; \alpha) \in C^1(G) \cap C(\bar{G})$. If f is such that $f(x) \neq 0, x \in \partial G_{[\alpha]}, \alpha \in [\alpha_1, \alpha_2]$, then $d\big(f(\cdot; \alpha_1), G_{[\alpha_1]}\big) = d\big(f(\cdot; \alpha_2), G_{[\alpha_2]}\big)$.

Chapter 16 – The Brouwer Degree

In the next chapter we'll use this to prove the Rabinowitz theorem.

Chapter 16 – The Brouwer Degree

Chapter 17 – The Rabinowitz Theorem

Now we will use the Brouwer degree to prove the Rabinowitz Theorem. The version of the theorem we will prove states

> Let $f(x; \alpha) = \Lambda(\alpha)x + h(x; \alpha)x$ with $h=o(1)$. Suppose that α_0 is an odd ordered root of $\det \Lambda(\alpha)$ (and thus a bifurcation point). Then the bifurcated solution, which exists due to the degree bifurcation theorem in Chapter 16, either is unbounded or connects to another bifurcation point, $(0; \alpha_1)$.

Suppose the theorem is not true. Then we have a bifurcated solution that is

i)   bounded, and
ii)   is bounded away from the zero solution.

Since the solution is bounded, we can find a set $G \subset \mathbb{R}^n$ and an interval $[a, b]$, $a < \alpha_0 < b$, such that the entire bifurcated solution curve is contained in $G \times [a, b]$. Since α_0 is an isolated root of $\det \Lambda(\alpha)$

The Rabinowitz Theorem

we can find $\delta > 0$ such that there are no other roots of $\det \Lambda(\alpha)$ in the interval $[\alpha_0 - \delta, \alpha_0 + \delta]$.

Let us describe the bifurcated curve by

$$(x(\tau); \alpha(\tau)) \qquad (17.1)$$

with

$$x(0) = \mathbf{0}, \alpha(0) = \alpha_0. \qquad (17.2)$$

By assumption the solution curve, except near $(0; \alpha_0)$, is bounded away from zero, so there is an $\varepsilon > 0$ such that if τ is such that $\alpha(\tau) \in [a, \alpha_0 - \delta] \cup [\alpha_0 + \delta, b]$ then

$$x(\tau) > \varepsilon. \qquad (17.3)$$

Remove the set $\{(x; \alpha): |x| < \varepsilon, |\alpha - \alpha_0| > \delta\}$. Also, since $(x(\tau); \alpha(\tau))$ is bounded away from any other solution curves, there is a $\mu > 0$ such that $(x(\tau); \alpha(\tau))$ stays at least μ away from all other solutions. Remove all points within μ of another solution.

The resulting set is shown schematically in Figure 17.1, with the horizontal axis the α-axis, and the vertical axis representing $\mathbb{R}^n$, x-space. This will be our set G.

The Rabinowitz Theorem

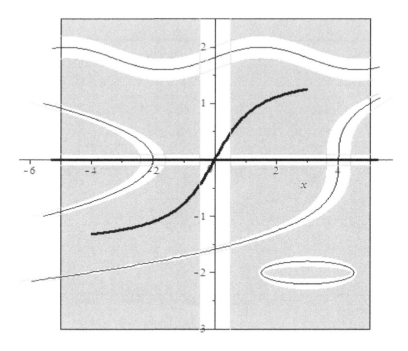

Figure 17.1

Now we will consider the degree on α-slices of this set. Consider the slice of the set at $\alpha = a$, $G_{[a]}$. A schematic of this is shown in Figure 17.2

The Rabinowitz Theorem

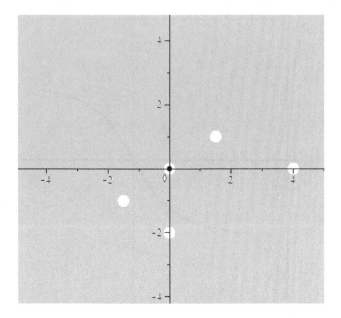

Figure 17.2

By construction, $f(x; a) = 0$ has no solutions in $G_{[a]}$, so $d\bigl(f(\,\cdot\,; a), G_{[a]}\bigr) = 0$.

Using the set Homotopy Theorem, we let α increase to $\alpha_0 - \delta$. As we approach $\alpha_0 - \delta$ from below we have the situation shown in Figure 17.3. By the Homotopy Theorem the degree is still equal to zero.

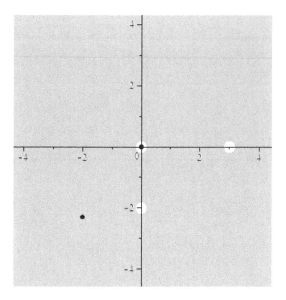

Figure 17.3

Now at $\alpha_0 - \delta$ we wish to add back the circle of radius ε centered at 0, as shown in Figure 17.4. This changes the degree, adding $signum(\det \Lambda(\alpha_0 - \delta))$, thus our degree is now ± 1:

$$d(f(\cdot\,;\alpha_0 - \delta), G_{[(\alpha_0-\delta)^+]} = d_1 = signum(\det \Lambda(\alpha_0 - \delta)) = \pm 1. \qquad (17.4)$$

The Rabinowitz Theorem

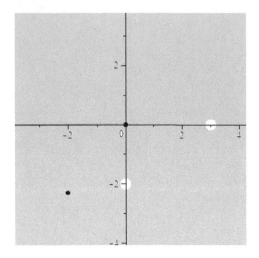

Figure 17.4

Again we use the Homotopy Theorem to increase α to $\alpha_0 + \delta$, as in Figure 17.5.

The Rabinowitz Theorem

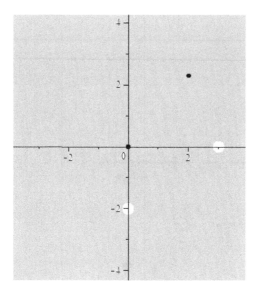

Figure 17.5

The degree has not changed. We now use the excision to remove the circle of radius ε around 0 again, as in Figure 17.6.

The Rabinowitz Theorem

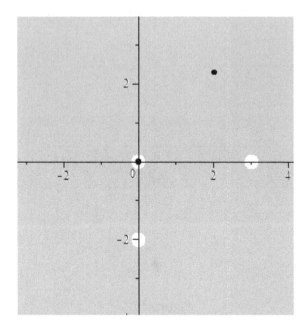

Figure 17.6

This changes the degree by subtracting $signum(\det \Lambda(\alpha_0 + \delta))$. However, $signum(\det \Lambda(\alpha_0 + \delta))$ is opposite in sign from $signum(\det \Lambda(\alpha_0 - \delta))$, since the root at α_0 was of odd multiplicity. Thus

$$d(f(\,\cdot\,;\alpha_0 + \delta), G_{[(\alpha_0+\delta)^+]} = d_1 - signum(\det \Lambda(\alpha_0 + \delta)) = \pm 2. \quad (17.5)$$

Finally, we homotopy to $\alpha = b$. The degree does not change, but the picture at $\alpha = b$ looks like Figure 17.7. There are no roots of f, so the degree must be zero.

The Rabinowitz Theorem

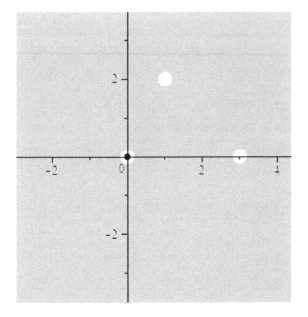

Figure 17.7

This contradicts our original assumption that $(x(\tau); \alpha(\tau))$ was bounded away from 0 and ∞. Thus we have proven the Rabinowitz Theorem.

Made in United States
Orlando, FL
17 January 2024

42582160R00114